ADVANCES IN APPLIED SCIENCE AND ENGINEERING

RESEARCH CHALLENGES IN SCIENCE, ENGINEERING AND TECHNOLOGY

Advances in Applied Science and Engineering

A. K. Haghi - Series Editor

Professor Emeritus of Engineering Sciences,
University of Guilan, Iran

Innovations and Challenges in Natural Products Chemistry: Research and Applications
A. K. Haghi (Editor)
2020. ISBN: 978-1-53618-650-5 (e-book)

Physical Chemistry and Its Interdisciplinary Applications
A. K. Haghi (Editor)
2020. ISBN: 978-1-53618-328-3 (e-book)

COVID-19 Pandemic: Questions, Answers and Hypotheses
A. K. Haghi (Editor)
2020. ISBN: 978-1-53618-565-2 (Softcover)
2020. ISBN: 978-1-53618-593-5 (e-book)

Innovations and Challenges in Modern Physical Chemistry: Research and Practices
A. K. Haghi (Editor)
2020. ISBN: 978-1-53618-181-4 (Hardcover)
2020. ISBN: 978-1-53618-235-4 (e-book)

Research Methodologies in Modern Chemical Technology: New Insights with Applications
A. K. Haghi (Editor)
2020. ISBN: 978-1-53618-562-1 (e-book)

ADVANCES IN APPLIED SCIENCE AND ENGINEERING

RESEARCH CHALLENGES IN SCIENCE, ENGINEERING AND TECHNOLOGY

S. KANNADHASAN
R. NAGARAJAN
AND
M. SHANMUGANANTHAM
EDITORS

nova
science publishers
New York

NOTICE TO THE READER

Library of Congress Cataloging-in-Publication Data

ISBN: 978-1-68507-008-3

Published by Nova Science Publishers, Inc. † New York

CONTENTS

Preface		**ix**
Chapter 1	Evaluation and Learning Methods of Artificial Intelligence *Neenu Ann Sunny*	**1**
Chapter 2	Computerised Medical Image Processing *V. V. R. Raman and B. Leelamani*	**21**
Chapter 3	A Review on Data Driven Based Discrimination Techniques between Faults and Power Swing for Hybrid Renewable Energy Systems *Swati S. Pachauri, Sonal Tripathi, Abdur Rahman Khan, Bhaumik Joshi, Siddarath Vats and Siddhi Vinayak Pandey*	**41**
Chapter 4	Antimicrobial Behaviour of Substituted Styryl 3-Nitro-4-Chlorophenyl Ketone *V. Usha, G. Thirunarayanan, V. Thangaraj, I. Muthuvel and S. Rajalakshmi*	**65**
Chapter 5	Computation of Congestion Cost in Transmission Lines *S. Chellam, M. Jajini and S. Kalyani*	**85**

Chapter 6	Human Computer Interaction for Detection of Hand Gesture Using 3D Convolutional Neural Network in Machine Learning Techniques *I. Varalakshmi, D. Mohana Priya and P. Mathivanan*	**95**
Chapter 7	Community Energy Storage: National and International Scenario - A Review *Gauri M. Karve, Kalyani M. Kurundkar and Geetanjali A. Vaidya*	**115**
Chapter 8	Modelling of Multi-Input Three Level DC-DC Converter *K. K. Saravanan and B. Rajeshkanna*	**135**
Chapter 9	E-Learning and Beyond: The Future of Technology in Education *K. V. Binoy Mathew and Maryelizabeth Tidiya Walarine*	**151**
Chapter 10	State of Charge Estimation Techniques Using the Filter-Based and Data-Driven Approach: A Review *Siddhi Vinayak Pandey and Soham Shah*	**163**
Chapter 11	A Superlative Set Theory for the Prediction of Diabetic Risk *M. Durgadevi*	**195**
Chapter 12	Investigation of Boost Converter Using Enhanced Dynamic Power Factor Rectification *R. Felshiya Rajakumari, S. Gopinath and S. Pragadeswaran*	**213**
Chapter 13	Hydrogen Fuel Cell for Automobile Industries: A Review *Nilay Poharkar, Ayushi Gupta, Akshay Srivastava, Jeet Patel and Siddhi Vinayak Pandey*	**225**

Chapter 14 MediSearch: A Framework for Keyword Search on Medical Terms Related to Air Pollution **257**
Bhimavarapu Usharani

Chapter 15 Automatic Hibiscus Leaf Disease Detection and Classification Using Unsupervised Learning Techniques **275**
Bhimavarapu Usharani

Chapter 16 Recent Trends in Solar Energy Generation, Challenges and Opportunities **287**
Guruprasath Rengarajan and Alagar Karthik

Chapter 17 Optimized Antireflection Coatings for Higher Efficiency ZnO Solar Cell **297**
S. Rajalakshmi, N. Sundaramurth and V. Usha

Chapter 18 Structural Properties of Fe Doped ZnO Nanoparticles **321**
S. Rajalakshmi, N. Sundaramurth and V. Usha

Editors' Contact Information **335**

Index **337**

PREFACE

This edited volume includes eighteen chapters and discusses various research challenges in science, engineering and technology. Topics discussed include learning methods of artificial intelligence, computerized medical image processing, human-computer interaction for detection of hand gestures, community energy storage, e-learning, prediction of diabetic risk, hydrogen fuel cells for automobiles, solar cells, and more.

Chapter 1 - This part of software engineering is worried about causing PCs to act like people. Man-made brainpower incorporates game playing, master frameworks, neural organizations, characteristic language, and mechanical technology. Presently, no PCs show full man-made consciousness (that is, can reproduce human conduct). The best advances have happened in the field of games playing. The best PC chess programs are presently equipped for beating people. Today, the most sweltering zone of man-made reasoning is neural organizations, which are demonstrating fruitful in various teaches, for example, voice acknowledgment and normal language handling. There are a few programming dialects that are known as AI dialects since they are utilized solely for AI applications. The two most regular are LISP and Prolog. Man-made brainpower is working a great deal in diminishing human exertion however with less development.

Chapter 2 - Introduction: Over the past 25 years, techniques to image brain structure and performance have offered investigators within the cognitive neurosciences and related fields unprecedented opportunities to review how human brain systems work and are connected. This new rehabilitation research and development arena is vital because a greater understanding of how and why brain systems remap within the service of rehabilitation will cause the development of better treatment. Objective: The principal objectives of this course are to provide an introduction to basic concepts and techniques for medical image processing and to promote interests for further study and research in medical imaging processing. Modern science also has been doing wonders within the surgical field. But, the right and proper diagnosis of diseases is that the primary necessity before the treatment. The rapid progress of life science and therefore the invention of varied medicines have benefited mankind and the whole civilization. Background: The functional neuroimaging methods have been developed and new structural neuroimaging techniques also have been added to the tool box of rehabilitation researchers. For example, diffusion tensor imaging (DTI) and related magnetic resonance (MR) techniques offer the power to assess human white matter pathways in vivo. Not only can these techniques be used to estimate the integrity of a given volume of white matter, but they also can be used to trace fiber tracts within the brain. This latter development is exciting because most of what we know about the connections of the human cortex has actually come from research on nonhuman primates, leaving questions especially about the phylogenetically newer portions of the cortex. In the rehabilitation arena, a far better understanding of how the brain's connections are damaged could help us to predict what treatments are best for various research subjects and, eventually, might be useful in selecting the simplest treatment strategies for individual patients. Because the newer functional and structural neuroimaging techniques have enormous implications for rehabilitation research and development, it is highly desirable that rehabilitation researchers be ready to evaluate the usefulness of the techniques for rehabilitation research and that the consumers of rehabilitation research. The chapter will consist of two main sections: (1) Because of the extraordinary versatility or MR techniques, the large number of MR techniques will be discussed first. (2) Subsequently, other functional neuroimaging techniques will be discussed, including: Positron Emission Tomography (PET), Magneto encephalography/Magnetic

Source Imaging (MEG/MSI), Near Infrared Spectroscopy (NIRS), Trans cranial Magnetic Stimulation (TMS), and Electroencephalography/ Evoked Potentials (EEG/EPs). For each imaging modality, the authors will give a brief explanation of the modality, its uses/potential uses in rehabilitation research, its strengths and limitations, and an example of research in the area. Conclusion: Computer-aided diagnostic processing has already become an important part of clinical routine. Accompanied by a rush of latest development of technology and use of varied imaging modalities, more challenges arise; for instance, the way to process and analyse a big volume of images in order that top quality information are often produced for disease diagnoses and treatment.

Chapter 3 - Due to the limitations of non-renewable resources in terms of energy-efficient and sustainability, the demand for renewable-based energy generating stations is increasing drastically. In order to balance the supply-demand chain, the hybrid grid-connected systems are being used. Presently, most of the hybrid grid-connected systems are replacing their conventional (non-renewable) stations with renewable energy-based generating power plants. With this advancement, the discrimination between faults and power swing has always been a challenging task for all grid connected systems. Sometimes, it leads to the false tripping command to distance protection systems utilized within a transmission line. For such associated issues, the accurate discrimination between faults and power swing can avoid the mal-operation of distance protection scheme and let the systems work more efficiently without any intervention. The present paper provides detailed insights about various discrimination techniques utilized in modern grid-connected systems using data-driven techniques, such as - Support Vector Machine (SVM), Artificial Neural Network (ANN), Neuro-Fuzzy techniques, Bayesian Classifier & Adaptive Wavelet technique, Linear Discriminant Analysis (LDS), and transformations approach. Also, the comparative analysis between conventional-based discrimination approach with modern data-driven-based techniques has been explicated along with future works to be carried out for better optimization.

Chapter 4 - This article describes the synthesis of substituted styryl 3-nitro-4-chloro phenyl ketones. Molecular formula, molecular weight, melting point, CHN analysis were determined. The existence of *s-cis* and *s-trans* conformers is confirmed by IR spectra. Presence of H_α, H_β, ^{13}C of CO, C_α, C_β has been confirmed by NMR spectra. The antibacterial and antifungal activity of all the compounds is determined by disc diffusion method. All the compounds have satisfactory antimicrobial activity.

Chapter 5 - Transmission Congestion Cost represents the cost associated with re-dispatching of schedules to avoid transmission line limit constraints violations. Re-dispatching of units cause the increase of total generation cost. This paper is an effort to rationalize and systematize the procedures for calculation of transmission congestion cost using tracing based power flow. The power flow in the transmission lines is traced using Power world simulator tool. The increase in real power generation will cause overload at transmission lines. This congested power flow in transmission is employed to determine the congestion cost. The congestion cost is calculated from congested power flow and cost per unit power flow in transmission lines. The power transfer limit has been taken as security constraint while dealing with congestion problem. The proposed methodology is illustrated on IEEE 14 bus test system.

Chapter 6 - Human–Computer Interaction (HCI) is a trending technology that creates superlative communication between human and computer. In this chapter, the authors discuss HCI with hand gesture detection using various machine learning techniques and algorithms. The results are also discussed by comparing various techniques for better understanding. Gesture-based interfaces are used in many applications such as Smartphone's, Televisions (TVs), video gaming, and so on. In this chapter, a novel method for fingertip detection and hand gesture recognition system using an RGB-D camera and 3D convolution neural network (3DCNN). The key frames extracted from static video summarization based on clustering method and 3D CNN algorithm are used as input to improve classification accuracy. The network of processing layers is used to distinguish the hand contour coordinates and detect the fingertip using KCA (K cosine Algorithm). This chapter demonstrates the accuracy and robustness of the interface by evaluating the hand gesture recognition system from camera images. The various existing algorithms such as DC-CNN

(Double channel-CNN), K-curvature method, scale invariant feature transform (SIFT) for hand gesture detection have been discussed along with the parameters and results obtained. The calibration of hand position and orientation aims at translating and rotating the hand images to neutral pose and these data are trained by CNN classifier. The authors provided a validation of the proposed method on recognizing human gestures which shows robust results with various hand positions, orientations and light conditions.

Chapter 7 - 'Community Energy Storage (CES)' means a distributed group of small units of energy storage systems (ESS) connected to the load side of transformers which serves a small residential or small commercial load. CES is a storage owned electric utility which is distributed and located near to end-users. The idea of CES is to dispatch and optimize energy storage resources. The potential benefits of dispensing the storage capacity rather than using one or a few bigger units can be significant which are backup power, flicker mitigation and integration of renewable energy sources (RES) at local level, while at grid level, benefits of CES are ancillary services, load leveling and power factor correction. The CES can be of various types such as pumped hydro, compressed air, flywheel or battery storage. This work focuses on status and types of CES in India and in the world with their specific characteristics and capacities. The chapter is concluded with policy decisions regarding CES at few developed countries and its need in the other developing countries.

Chapter 8 - A multi-input DC-DC boost converter is proposed to obtain power from several input sources and to supply the regulated output voltage to the load from the sources. The structure of Multiple-Input Converter (MIC) is simpler compared to the use of several single-input converters for each source. As the conventional source of energy is rapidly depleting and the cost of energy is rising.in the multi-input converters offer a cost effective solution in applications which requires multiple input sources such as fuel cell vehicles and renewable energy systems. The basic idea is to integrate a number of converters in either input dc/dc conversion stage or in isolation stage, in addition to commonly shared output stage. There are various multi-input topologies proposed in literature based on non-isolate and isolated structures. In, a non-isolated multi-input buck/boost converter sharing the same switch has been proposed. For multi-input converters provide galvanic isolation, majority of the work has been devoted to phase-shifted full-bridge converters. In, two identical current fed full-bridge converters are connected in parallel to interface a common output rectifying diode-bridge. Similar concept has been applied using individual output rectification stages but sharing an integrated transformer in two current fed bridges are controlled in a phase-shifted manner. The parallel connections of multi-input cells are first identified as pulsating voltage-source cell and pulsating current-source cell, and various topological combinations have been presented.

Chapter 9 - The era of knowledge explosion forces us to be lifelong learners to be abreast in our chosen field. Advancements in information and communication technology have revolutionized technology-enhanced learning. The dawn of the twenty-first century has opened to us the vast world of online learning. Online learning methods have taken the place of traditional learning methods during the pandemic. Many online platforms have already put their footprints in the current educational scenario. Multidimensional learning methods are blended together to give a much better outcome than single-dimensional learning methods. Innovative learning methods have won the hearts of students. They have the potentiality to assist in achieving the comprehensive development of a student by bringing them out into the realm of practicability. The flipped classroom, Escape rooms, Gamification, M-learning, Immersive Virtual Reality, Competency-based education, Digital Problem Based Learning are some examples of such innovative learning methods.

Chapter 10 - The demand of batteries in electrical systems is increasing significantly. Many electrical vehicles (automobiles), various power plants and space satellites are using different types of batteries based on their applications. Among these, the quantification of State of Charge (SOC) is the consequential parameters for all the battery packs. The SOC represents available battery power in authentic time domain, which gives the information to users about how much time it can provide a potency to the loads. This paper describes various methods to estimate SOC in order to predict remaining battery energy in real time domain. However, there has been significant research on the SOC estimation techniques, out of which the most broadly used and highly accurate methods are described in this paper. The filter-based technologies such as - Kalman Filter (KF), Extended Kalman Filter (EKF), Unscented Kalman Filter (UKF)/Sigma Point Kalman Filter (SPKF) and various data driven techniques as - Support Vector Machines (SVMs), various Neural Networks such as - NARX NNs, NIO NNs, RNNs and DNNs are being discussed in detail. The Reinforcement Learning approach; which is one of the widely used Hybrid-estimation method for the accurate estimation of SOC has also been deliberated. The paper gives the literature survey and elucidates these technologies in considerable detail and culminates the future research requires in this field along with many different efficient estimation techniques for several types of linear and nonlinear battery loads.

Chapter 11 - In an actual risk-assessment process enormous classification and prediction algorithms have been developed in the field of data mining, among which only a few have employed mathematical predictions. This chapter introduces an innovative mathematical prediction model to determine the risk level of diabetes by considering different factors of diabetic risk such as Glycosylated Haemoglobin (HbA1c), Plasma Glucose-Fasting (PG-F), Plasma Glucose-Post Prandial (PG-PP) and age with reference to fuzzy set t heory and soft set theory. The real time diabetic dataset from a multispecialty diabetic clinic was used for risk analysis. The predictive ability of the superlative set theory was compared with the Grey prediction, Rough Set Theory and Fuzzy Set theory. The Superlative Set theory has reduced the unnecessary, irrelevant and redundant features in the datasets and has produced superior forecasting results in terms of accurate diabetic risk prediction.

Chapter 12 - In this work, boost converter power factor correction is analyzed. This work is discussed and design with a DC-DC boost converter and a single phase rectifier with improved power factor boost converter technique. The normal DC-DC boost converter is intend and the parameters such as ripple, input current, output voltage is executed and then power factor is calculated by interfacing the boost converter with rectifier. By designing the input technique the power factor will be improved. Proactive power factor correction and active power factor correction for the boost converter is discussed in this paper. By using two power factor techniques active power factor correction which gives the best result because the boost converter is coupled with rectifier circuit. Boost converter is one of the best methods to implement for the power factor value from the input sinusoidal AC voltage waveform.

Chapter 13 - Many automobile industries are moving towards the inclusion of renewable based energy generating resources due to the pollutions causes with nonrenewable (gasoline) based vehicles. Among them, electric and fuel cell based vehicles are leading in field of automobile industries Electric vehicles may not be an effective alternative for conventional gasoline conveyances due to its dependency nature on the way of electricity engenderment. The main aim of automakers is to minimize the exhaust emission and fuel consumption with optimizing the range, power and overall efficiency. In this paper, the rudimentary working principle of fuel cell along with its types, different kind of membranes and sundry comparisons such as - fuel cell electric vehicle with battery operated vehicle is elucidated in detail. This paper also present various kind of challenges and its solutions which required during the astute and efficient development of hydrogen fuel cell based infrastructure.

Chapter 14 - Air pollution is one of the major environmental issue of the developed and developing countries throughout the world. Around 90 percent people in the world are breathing the contaminated air. The universalecological pollution is assessedas aworld-wide public health concern with numerousissues. Amongst all the clinical search queries on the Internet, air pollution associated health issue data is one of the utmostwidespreadissues that originates much of the society's attention, specifically on the globe.There is the need of similarity-based technique to create word interpretations and to assess clinical terms. This paper discusses work on keyword search a kind of "medical keyword search" which can help users to find medical terms on the internet. MediSearch combines Google web API with a hierarchy visualization technique. It permits the user to perform keyword search on medical terms using medical related files and to outwardly examine the documents to check their pertinence and get the natural remedies for air pollution related health issues.

Chapter 15 - Hibiscus is a fantastic herb and in Ayurveda it is one of the most renowned herbs that have extraordinary healing properties.

Hibiscus is rich in vitamin C, flavonoids, amino acids, mucilage fibre, moisture content and antioxidants. Hibiscus can help with weight loss, cancer treatment, bacterial infections, fever, high blood pressure, lower body temperature, treat heart and nerve diseases. Automatic leaf disease detection is an essential task. Image Processing is one of the popular techniques for the plant leaf disease detection and categorization. In this paper, the diseased leaf is identified by Concurrent K-Means Clustering algorithm and then features were extracted. Finally, Reweighted KNN Linear Classification algorithms have been used to detect the diseased leaves categories.

Chapter 16 - Solar energy innovations have seen phenomenal development in recent years. This strong involvement in these innovations has been encouraged and maintained by the realization of technical advances, rising public knowledge of environmental problems, the economic situation, and the number of policy instruments. Although the cost of solar-generated electricity is still costly and the power from renewable energies, like solar, is still low energy, there has been no large-scale production of renewable resources and generation providers are still least involved in purchasing power from renewable sources. This paper gives an analysis of the technological, economic and policy implications of the production of solar energy.

Chapter 17 - Dye-sensitized solar cells (DSSCs) have been widely paid attention due to their great potential for low cost energy conversion devices. However, DSSCs show low power conversion efficiency (PCE) because of some efficiency losses, such as reflection of light by a substrate, charge recombination, carrier trapping in a material and surface power dispersion due to the resistance of cell (material, metal, electrode, etc). Approaches to improve the power conversion efficiency of DSSCs are to reduce these losses. Reduction of light reflection by using three layer anti-reflection coatings is one of the effective approaches. These three layers are expected to reduce the reflection and increase the transmission of light at the glass surface. This leads to an increase of the total flux of photons to photocurrent through the electrode and short circuit current density (Jsc).

Chapter 18 - The brief introduction to nanotechnology, it is important to discuss about the Ferric nano materials and its application of various fields. Moreover, the highlights of Fe doped ZnO nanoparticles and its applications were explained under suitable conditions. The synthesis method adopted for the preparation of ZnO and Fe doped ZnO nanoparticles are discussed. The detailed experimental techniques used to characterize the synthesized Fe doped ZnO nanoparticles are explained well and the results elaborate the various studies such as XRD, FTIR, VSM, which have been carried out of synthesized nanoparticles.

In: Research Challenges in Science … ISBN: 978-1-68507-008-3
Editors: S. Kannadhasan et al.

Chapter 1

EVALUATION AND LEARNING METHODS OF ARTIFICIAL INTELLIGENCE

Neenu Ann Sunny*

Department of Computer Applications, Saintgits College of Applied Sciences Pathamuttom, Kottayam, India

ABSTRACT

This part of software engineering is worried about causing PCs to act like people. Man-made brainpower incorporates game playing, master frameworks, neural organizations, characteristic language, and mechanical technology. Presently, no PCs show full man-made consciousness (that is, can reproduce human conduct). The best advances have happened in the field of games playing. The best PC chess programs are presently equipped for beating people. Today, the most sweltering zone of man-made reasoning is neural organizations, which are demonstrating fruitful in various teaches, for example, voice acknowledgment and normal language handling. There are a few programming dialects that are known as AI dialects since they are utilized solely for AI applications. The two most regular are LISP and Prolog. Man-made brainpower is working a great deal in diminishing human exertion however with less development.

---

*Corresponding Author's E-mail: neenuanns@gmail.com.

Keywords: data, mining, epistemology, ontology, heuristics, optimization

INTRODUCTION

Man-made brainpower is characterized ascreating computer projects to take care of complex issues by human thinking measures. It is that part of software engineering that reviews and creates astute machines and programming. The field was established on the case that a focal property of people, knowledge - the intelligence of Homo sapiens - can be so exactly portrayed that it very well may be mimicked by a machine. This raises philosophical issues about the idea of the psyche and the morals of making fake creatures, issues which have been tended to by fantasy, fiction, and reasoning since vestige. Man-made brainpower has been the subject of colossal hopefulness however has likewise endured shocking difficulties. Today it has become a fundamental piece of the innovation business, giving the hard work to a significant number of the most troublesome issues in software engineering.

The focal issues (or objectives) of AI research incorporatethinking, data arranging, learning, correspondence, observation, and the capacity to move and control objects. There are a huge number of instruments utilized in AI, including variants of search and numerical improvement, rationale, strategies dependent on likelihood and financial aspects, and numerous others.

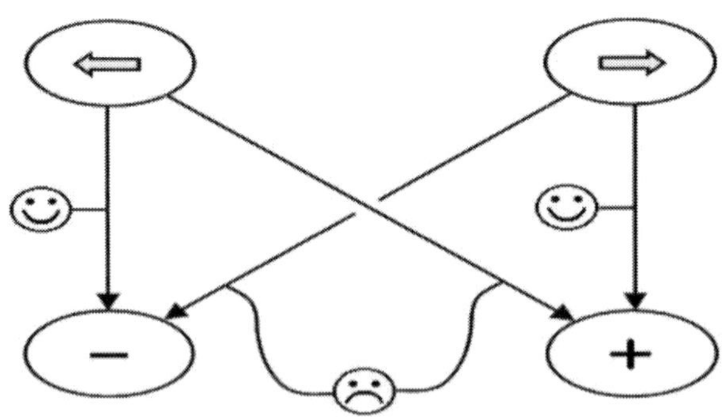

Figure 1. Artificial intelligence.

LITERATURE REVIEW

The writing audit ought to exhibit that the writer has decent information on the exploration territory. Writing survey ordinarily possesses a couple of entries in the presentation area. An elegantly composed writing audit ought to give a basic examination of past investigations identified with the ebb and flow research region as opposed to a straightforward outline of earlier works. The creator shouldn't avoid calling attention to the deficiencies of past works. Notwithstanding, scrutinizing other's work with no premise can debilitate your paper. This is an ideal spot to coin your examination address and legitimize the requirement for such an investigation. It is likewise worth bringing up towards the finish of the survey that your investigation is exceptional and there is no immediate writing tending to this issue. Add a couple of sentences about the meaning of your examination and how this will enhance the assortment of information.

HISTORY OF ARTIFICIAL INTELLIGENCE

Man-made consciousness is certainly not another word and not another innovation for scientists. This innovation is a lot more established than you would envision. Indeed, even there are legends of Mechanical men in Ancient Greek and Egyptian Myths. Following are a few milestones in the history of AI which define the journey from the AI generation to date development.

Maturation of Artificial Intelligence (1943-1952)

- The year 1943: The primary work which is currently perceived as AI was finished by Warren McCulloch and Walter pits in 1943. They proposed a model of counterfeit neurons
- The year 1949: Donald Hebb exhibited a refreshing standard for adjusting the association strength between neurons. His standard is currently called Hebbian learning.

- The year 1950: Alan Turing was an English mathematician and spearheaded Machine learning in 1950. Alan Turing distributes "Processing Machinery and Intelligence" in which he proposed a test. The test can check the machine's capacity to display canny conduct identical to human insight, called a Turing test. The birth of Artificial Intelligence (1952-1956)
- The year 1955: An Allen Newell and Herbert A. Simon made the "principal computerized reasoning program" Which was named as
- "Rationale Theorist." This program had demonstrated 38 of 52 Mathematics hypotheses, and discover new and more exquisite evidence for certain hypotheses.
- The year 1956: "rtificial Intelligence" first received by AmericanComputer researcher JohnMcCarthy at the Dartmouth Conference. Unexpectedly, AI begat as a scholastic field.

Around then undeniable level codes, for example, FORTRAN, LISP, or COBOL were created.Also, the energy for AI was high around then.

The Golden Years-Early Enthusiasm (1956-1974)

- The year 1966: The scientists accentuated creating calculations that can take care of numerical issues. Joseph Weizenbaum made the first chatbot in 1966, which was named ELIZA.
- The year 1972: The primary canny humanoid robot was inherent Japan which was named WABOT-1.

The First AI Winter (1974-1980)

- The span between the years 1974 to 1980 was the principal AI winter term. Man-made intelligence winter alludes to the time frame where PC researchers managed a serious deficiency of subsidizing from the government for AI exploration.

- During AI winters, an interest in exposure to man-made consciousness was diminished.

A Boom of AI (1980-1987)

- The Year 1980: After AI winter length, AI returned with "Master System." Master frameworks were modified that imitate the dynamic capacity of a human master.
- In the Year 1980, the main public meeting of the AmericanAssociation of Artificial Intelligence was held at Stanford University.

The Second AI Winter (1987-1993)

- The length between the years 1987 to 1993 was the subsequent AI Winter span.
- Again Investors and the government halted subsidizing AI research because of significant expense however not an effective outcome. The master framework, for example, XCON was exceptionally practical.

The Emergence of Intelligent Agents (1993-2011)

- The year 1997: In the year 1997, IBM Deep Blue beats world chess champion, Gary Kasparov, and turned into the principal PC to beat a world chess champion.
- The year 2002: unexpectedly, AI entered the home as Roomba, a vacuum cleaner.
- The year 2006: Computer-based intelligence came into the Business world till the year 2006. Organizations like Facebook, Twitter, and Netflix additionally began utilizing AI.

Deep Learning, Big Data, and Artificial General Intelligence (2011-Present)

- The year 2011: In the year 2011, IBM's Watson won risk, a test show, where it needed to tackle the unpredictable inquiries just as conundrums. Watson had demonstrated that it could comprehend common language and can tackle interesting inquiries rapidly.
- The year 2012: Google has dispatched an Android application highlight "Google now," which had the option to give data to the client as an expectation.
- The year 2014: In the year 2014, Chatbot "Eugene Goostman" won an opposition in the scandalous "Turing test."
- The year 2018: The "Undertaking Debater" from IBM bantered on complex subjects with two expert debaters and performed incredibly well.
- Google has shown an AI program "Duplex" which was a remote helper and which had taken stylish arrangement accessible as needs are, and woman on the opposite side didn't see that she was conversing with the machine

Presently AI has created to a surprising level. The idea of Deep learning, huge information, and information science are presently moving like a blast. These days organizations like Google, Facebook, IBM, and Amazon are working with AI and making astonishing gadgets. The fate of Artificial Intelligence is moving and will accompany high insight.

Components of AI

The User Interface

The UI is the methods forarticulation between a client and the master frameworks problem-solving measures. A decent master framework isn't helpful if not it has a valuable interface. It must have the option to affirm the headings in a structure that the client enters. It ought to decipher the appropriate responses, shaped by the framework.

The Information Base

It stores the real factors and rulesabout a demanding issue area. It makes these open to the induction motor in a structure that it can utilize. The specifics might be in a type of foundation all together incorporated into the framework. The principles incorporate both the assembling decides that worry to the territory of master framework and the heuristic and dependableguidelines that are given by the region master to make the framework discover arrangements.

The Shell or interface Engine

The interface motor is the program thatfinds the reasonable data in the database, and derives new data by applying steady handling and insightful procedures.

Branches of Artificial Intelligence

Genetic Programming

Hereditary programming rotatesaround that works that get projects to do correct undertakings and take care of definite issues. This is finished by 'cover together' irregular drawl programs and picking between a huge number of age. By importance, hereditary writing computer programs is a creative algorithm-based technique that can attract its parentage to the natural development that involved the quest for PC programs that performed client characterized undertakings. The world can express gratitude toward JOHN KOZA and his gathering of specialists in 'Computerized reasoning' for the technique for hereditary programming.

Ontology

Metaphysics is a part of man-made reasoning that is on edge with the investigation of different sorts of articles. It is a bunch of ideas that are officially spoken to inside a territory. The bar is then unfortunate on the relationship between the ideas in their precise spaces. Cosmology can likewise be utilized to cause about substance inside a demanding zone and clarify the demanding region in detail.

Epistemology

Epistemology frets about the investigation of the data that can show accommodating in the investigation of the settling of the damages that illness the world. Epistemology has figured out how to cut for itself a spot in counterfeit designing. As a part of computerized reasoning, be that as it may, epistemology centers on addressing four focus questions: What is data? How is data procured? What do individuals know? How would weunderstand what we know? The term 'Epistemology' wasfirst brought into the English language by the Scottish savant James Frederick Ferrier.

Heuristics

It is a part of man-made reasoning that frets about the experience-based procedures for taking care of issues, learning, and disclosure. Heuristic strategies and procedures are generally concerning quickly and capably finding an ideal response for a particular issue. Heuristics, typically talk, is about judgment numerous or careful answers for an accurate issue.

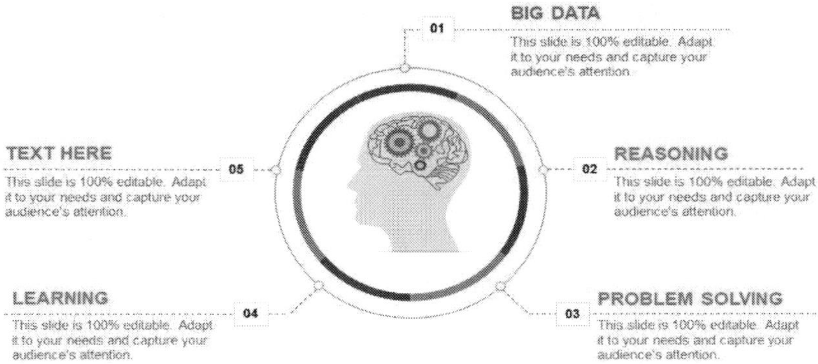

Figure 2. 5 Benefits of Artificial Intelligence.

Benefits of Artificial Intelligence

Search and Optimization

Numerous issues in AI can befathomed in principle by insightfully looking through numerous potential arrangements that are thinking can be diminished to playing out a pursuit. For instance, consistent evidence can be seen as looking for a way that leads from premises to ends, where each progression is the use of an induction rule. Arranging calculations search through trees of objectives and sub-objectives, endeavoring to discover a way to an objective, a cycle called implies closes examination. Advanced mechanics calculations for moving appendages and getting a handleon articles utilizeneighborhood look in arrangement space. Basic comprehensive ventures are seldom adequate for most true issues: the hunt space (the number of spots to look) rapidly develops to galactic numbers. The arrangement, for some, issues, is to utilize "heuristics" or "dependable guidelines" that take out decisions that are probably not going to lead to the objective (called "pruning the pursuit tree").

Heuristics flexibly the program with a "best theory" for the way in which the arrangement lies. These calculations can be envisioned as visually impaired slope climbing: we start the hunt at an arbitrary point on the scene, and afterward, by bounces or steps, we continue moving our estimate tough, until we arrive at the top. Developmental calculation utilizes a type of advancement search. For instance, they may start with a populace of life forms (the conjectures) and afterward permit them to change and recombine, choosing simply the fittest to endure every age (refining the estimates).

Logic

The rationale is utilized for data portrayal and critical thinking, yet it tends to be applied to different issues also. A few unique types of rationale are utilized in AI Research. The propositional or sentential rationale is the rationale of proclamations that can be valid or bogus. The first-request rationale additionally permits the utilization of quantifiers and predicates and can communicate realities about items, their properties, and their relations with one another. The fluffy rationale is an adaptation of the First-request rationale which permits the reality of an assertion to be spoken to as an incentive somewhere in the range of 0 and 1, as opposed to just evident (1) or bogus (0). Fluffy frameworks can be utilized for unsure Reasoning and have been broadly utilized in present-day mechanical and customer item control Systems. Abstract rationale models vulnerability in an alternate and more expressway than Fuzzy rationale.

Default rationales, non-monotonicrationales and circumscription are types of rationale intended to help with Default thinking and the capability issue. A few augmentations of rationale have been intended to deal with explicit spaces of data, for example, depiction rationales, circumstance analytics, occasion Calculus, and familiar math (for speaking to occasions and time), causal analytics; conviction math, and modular rationales.

Probabilistic Methods for Uncertain Reasoning

Numerous issues in AI (in thinking, arranging, learning, observation, and mechanical technology) require the specialist to work with inadequate or questionable data. Artificial intelligence specialists have formulated various amazing assets to take care of these issues utilizing techniques from likelihood hypothesis and financial aspects. Bayesian organizations are an overall instrument that can be utilized for countless issues: thinking (utilizing the Bayesian deduction calculation), picking up (utilizing the desire amplification calculation), and arranging (utilizing choice organizations), and discernment (utilizing dynamic Bayesian organizations). Probabilistic calculations can likewise be utilized for separating, forecast, smoothing, and discovering clarifications for surges of information, assisting recognition frameworks with breaking down cycles that happen over the long run.

A critical idea from the study of financial matters is "utility": a proportion of how significant something is to a savvy specialist. Exact numerical devices have been built up that dissect how a specialist can settle on decisions and plan, utilizing choice hypothesis, choice investigation, data esteem hypothesis. These apparatuses incorporate models, for example, dynamic choice organizations, game hypothesis, and component plan.

Classifiers and Statistical Learning Methods

The easiest AI applications can bepartitioned into two kinds: classifiers ("on the off chance that glossy, at that point jewel") and regulators ("if gleaming, at that point get"). Regulators do anyway additionally order conditions before surmising activities, and hence characterization frames a focal piece of numerous AI frameworks. Classifiers are capacities that utilization design coordinating to decide the nearest coordinate. They can be tuned by models, making them alluring for use in AI. These models are known as perceptions or examples.

In managed learning, each example has a place with a certain predefinedclass. A class can be viewed as a choice that must be made.

All the perceptions joined with their group names are known as an informational index. At the point when a groundbreaking perception is gotten, that perception is characterized dependent on experience. A classifier can be prepared differently; there are numerous measurable and AI draws near. The most generally utilized classifiers are the neural organization, portion techniques, for example, the help vector machine,k-closest neighbor calculation, Gaussian blend model, NAIVE BAYES classifier, and choice tree. The presentation of these classifiers has been thought about over a wide scope of assignments. Classifier execution relies enormously upon the attributes of the information to be grouped. There is no single classifier that works best on completely given issues. This is additionally alluded to as the "no free lunch" hypothesis. Deciding a reasonable classifier for a given issue is, even more, workmanship than science.

Neural Networks

A neural organization is aninterconnected gathering of hubs likened to the huge organization of neurons in the human mind. The investigation of fake neural organizations started in the decade before the field of AI research was established, in crafted by Walter Pitts and Warren McCullough. Early specialists were Frank Rosenblatt, who imagined the recognition, and PAUL WERBOS who built up the backproliferation calculation

Optimisation

Function Optimization

This is the employment of finding the best arrangement of boundaries of a capacity. There are numerous straightforward methods of doing this, tallying slope climbing. Figuratively, hillclimbing checks out the current situation for a higher position and moves to it. The top is reached if there is no higher position. This strategy is very guileless or less successful and can prompt discovering problematic arrangements (called neighborhood maxima).

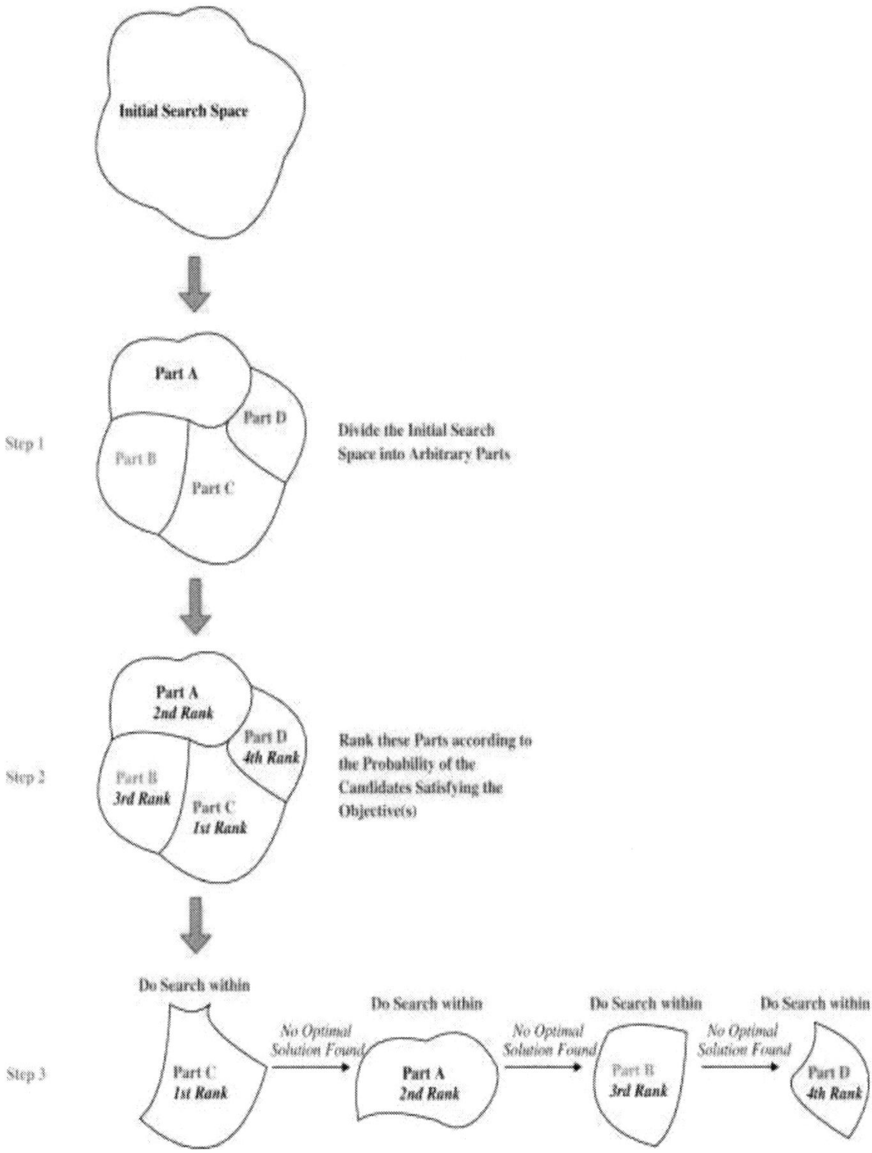

Figure 3.Terms of artificial intelligence.

Enhancement Capabilities arelikewise given by Genetic Algorithms, by impersonating the Process of advancement (as per Darwin's hypothesis) and natural selection. The best arrangements are matched off together to create better posterity arrangements. This strategy experiences less difficulty with nearby maxima, however, there are still no assurances of finding the ideal arrangement.

Planning

Arranging includes finding athe succession of activities that can lead from the present status to the objective state. This is generally done in various leveled ways: by and large, plans are exceptionally organized first, and the subtleties are worked out later. This is a more productive methodology. The blemished world is a serious issues that arranging needs to rival. With amazing conditions, a straightforward inquiry can be executed, and if an outcome is discovered, it will be conceivable practically speaking. It isn't generally conceivable to get wanted outcomes so, not work out or doesn't have the foreseen results then the plans don't work out.

Machine Learning

AI is getting progressively well known, and uniformly indispensable. Individuals feel that it is a lot more straightforward to get a machine to take in something from realities, as opposed to the need to invest energy showing it expressly. The central point is the nature of the learning calculation.

Neural Network

Fake Neural Networks, otherwise called Neural Networks (NN), are demonstrated on the human cerebrum. In this, the inward structure is comprised of few fake neurons, which recommends that the data educated isn't ideal. There is, be that as it may, the advantage of having the option, to sum up, for example, work with data that it didn't run over during its preparation. It is significant that how well the organization was planned and prepared on which depends on how well it can sum up which thusly infers how well it performs. Thusly, a great deal of examination is done on the strategies to make sure about great speculation.

Inductive Programming

Given just the results of acapacity (a restricted measure of them), inductive programming attempts to compose the meaning of the program that made those outcomes. This is pretty much effective relying upon the number of model outcomes were given, and how complex the capacity is. Right now, some inductive programming calculations can learn straightforward rationale programs, even recursively characterized. More troublesome projects won't just demonstrate hard to adapt yet in addition testing to execute this cycle to genuine information as opposed to PCdelivered capacities.

Decision Tree Learning

A choice tree is a structure that grants learning ofviewpoint (for example positive or negative) about items set up on their qualities (length, shading). Given a progression of models, the learning calculation can build a choice tree that will be equipped for arranging new models. On the off chance that the new models are controlled appropriately, nothing is finished. Something else, the structure of the tree is customized until the specific results are shown. The test is to get the calculation to excel on enormous arrangements of information, taking care of deficiencies in qualities (clamor), and deciding the best attack of the tree to the preparation and breaking down information.

Data Mining

This is the course of gettingaccommodating guidelines from an enormous arrangement of information. At the point when patterns are considered, their root should be distinguished, and a standard expressing their relationship should be set up. Around there, the test is having the option to deal with a great deal of data effectively, and disregard the potential blunders.

Bayesian Networks

Bayesian Networks speak to the connection between factors. This is known as restrictive reliance: the condition of a variable may depend on numerous others. This can be indicated as a chart, and there's a shrewd calculation to figure the likelihood of obscure outcomes given existing data. Admitting, one regular grumbling against this methodology identifies with the plan; it tends to be to show such organizations. Accordingly, learning the arrangement.

Comparison Results and Discussion

The outcomes and conversation segments are one of the provoking areas to compose. It is imperative to design thissegment cautiously as it might contain a lot of logical information that should be introduced in an unmistakable and succinct style. The reason for a Results segment is to introduce the vital aftereffects of your exploration. Results and conversations can either be consolidated into one area or coordinated as independent segments relying upon the necessities of the diary to which you are presenting your exploration paper. Use subsections and subheadings to improve comprehensibility and clearness. Number all tables and figures with expressive titles. Present your outcomes as figures and tables and direct the peruser toward applicable things while examining the outcomes. This part should feature critical or intriguing discoveries alongside P esteems for factual tests. Make certain to incorporate negative outcomes and feature likely impediments of the paper. You will be censured by the commentators if you don't talk about the weaknesses of your examination. This regularly compensates for an extraordinary conversation area, so don't be hesitant to feature them.

CONCLUSION

As of recently, we have examinedabout the critical highlights of computerized reasoning for example its advantages, advances, its exact, and a decent definition. Presently we can say that making a machine or state robot isn't as simple as an ABC. It is hard to make a machine like people which can show feelings or think like people in various conditions. Presently we have acknowledged that man-made reasoning is the investigation of how to make things that can precisely work as people do. It is how we think reasonably, act astutely, think like people, work like people. We realize that through man-made reasoning, even PC has crushed humans in chess. So we can say that coming to so far has not gone waste, by one way or another, it is contributing towards the headway in theArtificial knowledge. As of now, there is noPC indicating full computerized reasoning, however, the course of making machines such as ourselves is on its way.

FUTURE SCOPE

It is difficult to foresee the fateof artificial insight. Man-made consciousness in the '90s was centered pretty much on upgrading human conditions. However, is that the main objective later on? Exploration is fixated on developing human-like machines or robots. This is because researchers are worried about human insight and are awestruck by attempting to duplicate it. On the off chance that machines begin accomplishing the work done by people, at that point, the function of people will change. The difficult work of analysts may take care of them some time or another and we will discover our work done by machines and a robot strolling with us.

REFERENCES

[1] *Research Paper on Artificial Intelligence.* Pooja Agarwal, Pooja Yadav, Neelam Sharma, Ruchika Uniyal, SwatiSharma (students, Department of computer sciences & Eng.).

[2] Cadena, Cesar; Carlone, Luca; Carrillo, Henry; Latif, Yasir; Scaramuzza, Davide; Neira, Jose; Reid, Ian; Leonard, John J.(December 2016). "Past, Present, and Future of SimultaneousLocalization and Mapping: Toward the Robust-Perception Age." *IEEE Transactions on Robotics*. 32 (6):1309-1332. arXiv:1606.05830. Bibcode: 2016arXiv160605830C. doi:10.1109/TRO.2016.2624754. S2CID2596787.

[3] Moravec 1988, p. 15.

[4] Chan, Szu Ping (15 November 2015).*This is what will happen when robots take over the world.* Archived from the original on 24 April 2018. Retrieved 23 April 2018.

[5] "IKEA furniture and the limits of AI." *The Economist*. 2018. Archived from the original on 24 April 2018. Retrieved 24 April 2018.

[6] S.Kannadhasan, G.Karthikeyan and V.Sethupathi, A Graph Theory Based Energy Efficient Clustering Techniques in Wireless Sensor Networks. *Information and Communication Technologies Organized by Noorul Islam University (ICT 2013) Nagercoil on 11-12 April 2013, Published for Conference Proceedings by IEEE Explore Digital Library* 978-1-4673-5758-6/13 @2013 IEEE.

[7] "Kismet." *MIT Artificial Intelligence Laboratory*, Humanoid Robotics Group. Archived from the original on 17 October 2014. Retrieved 25 October 2014.

[8] Thompson, Derek (2018). "What Jobs Will the Robots Take?." *The Atlantic*. Archived from the original on 24 April 2018. Retrieved 24 April 2018.

[9] S.Kannadhasan and R.Suresh, EMD Algorithm for Robust Image Watermarking. Recent Advances in Mechanical Engineering and Interdisciplinary Developments Organized by Ponjesly College of Engineering (ICRAMID 2014) Nagercoil on 7-8 March 2014, Published for *Advanced Materials Research* Vols.984-985 (2014) PP 1255-1260, ISSN No:1022-6680

[10] Scassellati, Brian (2002). "Theory of mind for a humanoid robot." *Autonomous Robots*. 12(1):13-24. Doi: 10.1023/A:1013298507114.S2CID1979315.

[11] Cao, Yongcan; Yu, Wenwu; Ren, Wei; Chen, Guanrong (February 2013). "An Overview of Recent Progress in the Study of Distributed Multi-Agent Coordination." *IEEE Transactions on Industrial Informatics.* 9 (1): 427-438. ArXiv:1207.3231.

[12] http://en.wikipedia.org/wiki/Artificial_intelligen.

[13] http://en.wikipedia.org/wiki/Artificial_intelligenCe#CITEREFRussellNorvig2003.

[14] www.cs.utexas.edu/users/ear/AILongForSeminar.ppt.

[15] www.amodit.com/en/AMODIT/Artificialintelligence-elements.aspx.

[16] www.militaryaiworks.com/tools.html.

[17] *Java T point* - Google Search.

In: Research Challenges in Science ... ISBN: 978-1-68507-008-3
Editors: S. Kannadhasan et al.

Chapter 2

COMPUTERISED MEDICAL IMAGE PROCESSING

V. V. R. Raman[1,*] and B. Leelamani[2,†]

[1]Department of Computer Science,
Aurora's Degree and PG College, Chikkadpally, Hyderabad, India
[2]Department of Biotechnology,
Aurora's Degree and PG College, Chikkadpally, Hyderabad, India

ABSTRACT

Introduction

Over the past 25 years, techniques to image brain structure and performance have offered investigators within the cognitive neurosciences and related fields unprecedented opportunities to review how human brain systems work and are connected. This new rehabilitation research and development arena is vital because a greater understanding of how and why brain systems remap within the service of rehabilitation will cause the development of better treatment.

* Corresponding Author's E-mail: ramanvvr@adc.edu.in.
† Corresponding Author's E-mail: leelamanib@adc.edu.in.

Objective

The principal objectives of this course are to provide an introduction to basic concepts and techniques for medical image processing and to promote interests for further study and research in medical imaging processing. Modern science also has been doing wonders within the surgical field. But, the right and proper diagnosis of diseases is that the primary necessity before the treatment. The rapid progress of life science and therefore the invention of varied medicines have benefited mankind and the whole civilization.

Background

The functional neuroimaging methods have been developed and new structural neuroimaging techniques also have been added to the tool box of rehabilitation researchers. For example, diffusion tensor imaging (DTI) and related magnetic resonance (MR) techniques offer the power to assess human white matter pathways in vivo. Not only can these techniques be used to estimate the integrity of a given volume of white matter, but they also can be used to trace fiber tracts within the brain. This latter development is exciting because most of what we know about the connections of the human cortex has actually come from research on nonhuman primates, leaving questions especially about the phylogenetically newer portions of the cortex. In the rehabilitation arena, a far better understanding of how the brain's connections are damaged could help us to predict what treatments are best for various research subjects and, eventually, might be useful in selecting the simplest treatment strategies for individual patients. Because the newer functional and structural neuroimaging techniques have enormous implications for rehabilitation research and development, it is highly desirable that rehabilitation researchers be ready to evaluate the usefulness of the techniques for rehabilitation research and that the consumers of rehabilitation research.

The chapter will consist of two main sections: (1) Because of the extraordinary versatility or MR techniques, the large number of MR techniques will be discussed first. (2) Subsequently, other functional neuroimaging techniques will be discussed, including: Positron Emission Tomography (PET), Magneto encephalography/ Magnetic Source Imaging (MEG/MSI), Near Infrared Spectroscopy (NIRS), Trans cranial Magnetic Stimulation (TMS), and Electroencephalography/Evoked Potentials (EEG/EPs). For each imaging modality, we will give a brief explanation of the modality, its uses/potential uses in rehabilitation research, its strengths and limitations, and an example of research in the area.

Conclusion

Computer-aided diagnostic processing has already become an important part of clinical routine. Accompanied by a rush of latest development of technology and use of varied imaging modalities, more challenges arise; for instance, the way to process and analyse a big volume of images in order that top quality information are often produced for disease diagnoses and treatment.

Keywords: medical image processing, cognitive neurosciences, bio imaging, neuroimaging

INTRODUCTION

Medical imaging, also referred to as radiology, is that the field of drugs during which medical professionals recreate various images of parts of the body for diagnostic or treatment purposes. Medical imaging procedures include non-invasive tests that allow doctors to diagnose injuries and diseases without being intrusive.

Biomedical image processing has experienced dramatic expansion, and has been an interdisciplinary research field attracting expertise from applied math, computer sciences, engineering, statistics, physics, biology and medicine. Computer-aided diagnostic processing has already become a crucial part of clinical routine.

Fundamental Steps in Image Processing

1. Image acquisition: to accumulate a digital image.
2. Image pre-processing: to enhance the image in ways in which increases the probabilities for fulfilment of the other processes.
3. Image segmentation: to partitions an input image into its constituent parts or objects.
4. Image representation: to convert the input file to a form suitable for computer processing.
5. Image description: to extract features that end in some quantitative information of interest or features that are basic for differentiating one class of objects from another.
6. Image recognition: to assign a label to an object supported the knowledge provided by its descriptors.
7. Image interpretation: to assign aiming to an ensemble of recognized objects.

Medical Imaging Procedures

Medical imaging may be a central part of the improved outcomes of modern medicine. Different types of medical imaging procedures include:

- Computerized tomography (CT scan)
- Magnetic resonance imaging (MRI)
- Nuclear Medicine Imaging (Positron Emission Tomography)
- Ultrasounds
- X-rays

CT Scanners

A computerised tomography scan, more commonly called a CT scan, can create an in depth image of the within of the body using x-rays and computers. It is different to an x-ray because it produced a cross-sectional image of the body, almost like an MRI, making them better at looking at soft tissue and more subtle parts of the image that an x-ray might not devour. They can be used to visualise bones, internal organs and blood vessels. Upper body, like the brain, neck, spine, chest and sinuses are commonly scanned. Monitoring is additionally a crucial use of CT scanners, as regular scans allow progress to be kept of any developing conditions, e.g., cancer.

Working

The patient lies on their back on a panel. This panel passes into the scanner, which rotates round the section of your body currently within the scanner. The patient must lie down till the scan can get a transparent image. Usually, the radiologist who is working the machine stands in another room to avoid the radiation, but can communicate with the patient via an intercom. The scan can take from 10 to 20 minutes, but results are available as soon as a computer has analysed the scans.

Advantages

- CT scans are fairly short - they only last about 10 - 20 minutes.
- The results are extremely fast compared to other sorts of scans.
- CT scans are painless, as they are non-invasive.

Disadvantages

- As with many of scans, your body is exposed to some radiation. The more of the patient's body that is scanned, the more radiation they are exposed to. However, they are designed to minimise radiation exposure.
- There is the possibility of an allergic reaction to the dye used.

MRI Scanners

A magnetic resonance imaging scan, more commonly referred to as an MRI scan, may be a detailed cross-sectional image of a part of the body. It is almost like a CT scan, but features a higher quality, so it is easier to ascertain differences in tissues, as shown within the picture below.

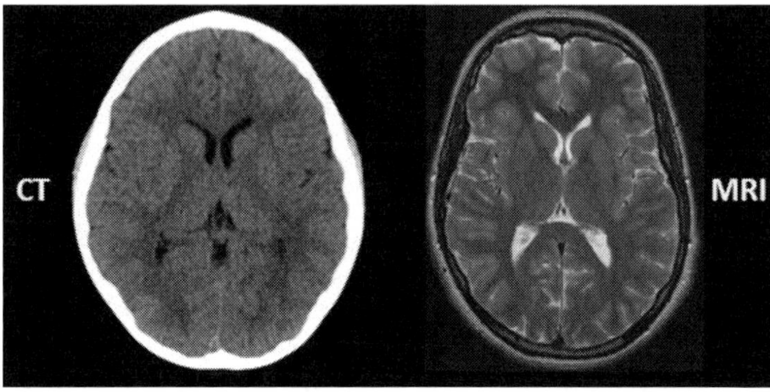

Figure 1. A CT scan vs. MRI scan.

MRIs are often used to get images of the brain and spinal cord, bones, the heart, blood vessels and different internal organs. The uses are almost like those of a CT scanner: diagnosis, getting more detail for treatment planning, and monitoring of on-going treatment.

Working

Unlike a CT scanner, an MRI surrounds the entire body. The patient is pushed into a thin tube; about 24 inches in diameter, and very strong magnets and radio waves are used to created detailed images. Similarly again to a CT scanner, the radiographer will substitute another room viewing the results and communicating via an intercom, but the MRI is far louder than a CT scanner. They can take from 15 to 90 minutes.

Advantages

- MRI scans are painless and safe, because the magnetic fields and radio waves haven't any known negative impact on the patient.

- They don't involve any x-ray radiation exposure, so are often employed by pregnant women and babies if necessary.

Disadvantages

- MRI scans enclose many of the body, so make people with claustrophobia uncomfortable.
- Metal cannot go inside of an MRI scanner, so people with certain implants like pacemakers cannot use them.

Nuclear Medicine Imaging (Positron Emission Tomography)

Positron emission tomography (PET) scans show brain processes by using the sugar glucose within the brain for instance where neurons are firing. Magnetic resonance imaging (MRI) scans to make a good discriminate among grey matter, white matter, and cerebrospinal fluid. A PET scan can create a 3D image of the inside of the body. They can be combined with CT and MRI scans to make a good clearer image to point out what's happening. They can even be focused on specific parts of the body, and show how well a part of the body is functioning. The image below shows how a PET scan and CT scan are often combined.

They are used to detect the progress of cancer, and can be used to get high resolution images of the brain. They are commonly used in people who have already been diagnosed with cancer, as they will clearly show how far a cancer has spread or skilled treatments like as chemotherapy. They are also utilized in the design of surgery, like brain or heart operations. Dementia can also be diagnosed with a PET scan, because it can show if the brain's normal function is modified.

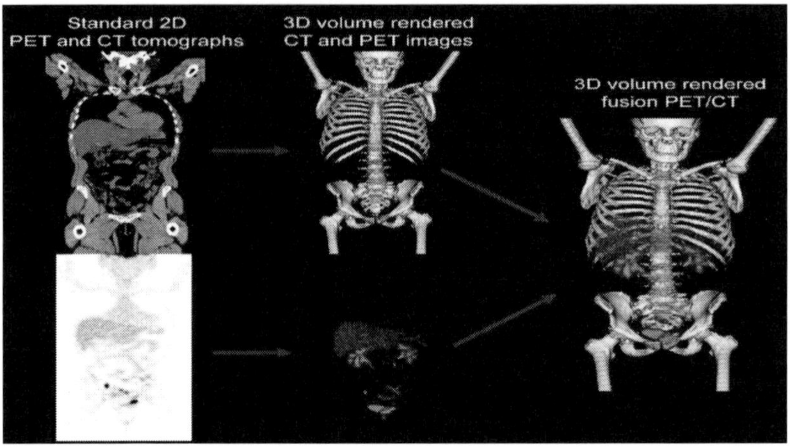

Figure 2. Combination of a CT scan and a PET scan.

Working

A radiotracer, usually fluorodeoxyglucose (FDG), is injected into your arm, and this gives off radiation. The PET scanner can detect this radiation when it collects in certain parts of your body. If there is an area where the FDG is not building up, then there is a particular body function which isn't working there. Cancerous cells use glucose at a faster rate than normal, and therefore by checking the FDG concentration, cancer can be identified and tracked in the body. The PET scanner machine looks almost like an MRI machine. The scan takes about 30 minutes.

Advantages

- The radiotracer fluorodeoxyglucose is used is similar to glucose, so the body treats it in a similar way.
- The scan only takes about 30 minutes.
- The PET scan can reveal cell level metabolic changes occurring in an organ or tissue, which a CT or MRI cannot.

Disadvantages

- The PET scan exposes you to radiation, which may lead to cancer. However, the amount is quite small. The radioactive tracer has a short half-life.
- Patients should avoid people who shouldn't be exposed to radiation, such as pregnant women, for a few hours after the scan.

Ultrasound

An ultrasound uses high-frequency waves to indicate what is within a region of the body. It is conjointly referred to as a sonogram. Ultrasounds can produce pictures of unborn kids in real time.

The most common use is to watch unborn babies, but they are also used in diagnosis and through bound procedures for guiding surgeons.

Working

The device features a probe that emits high-frequency sound waves. They bounce off of various components of the body, making echoes, and once these bounce back to the probe, it can also detect them. This can produce a live image on another scanner. The scan will last from 15 to 45 minutes. They can be done externally, internally or endoscopically.

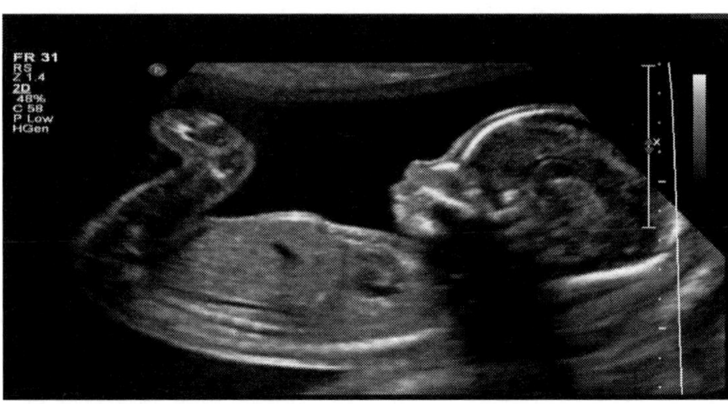

Figure 3. An ultrasound image of an unborn baby.

Advantages

- Usually there are no after-effects of ultrasound scans. This implies traditional activity will be resumed straight after.
- The results are seen in real time, thus there is no need to wait.

Disadvantages

- Some probe covers have latex, which may be a problem if the patient is allergic to latex.
- Endoscopic ultrasounds will cause an inflammatory disease or bloating, or in extreme cases, internal haemorrhage.

X-Rays

An x-ray could be a quite common procedure used to get pictures of within the body. It uses radiation within the x-ray a part of the electromagnetic spectrum. They are used to produce images of bones, sometimes to examine if and where there are breaks. They are conjointly employed by dentists and orthodontists to look at teeth. Bone tumours can also be seen on x-rays. They can be used to guide surgeons as they are operating. They can also be used to detect broken bones, and to arrange the simplest course of treatment for this.

Working

X-rays, which are a type of electromagnetic radiation invisible to humans, pass through the body. The energy is absorbed by different parts of the body at different rates, and a detector on the other side of the person can see what proportion was absorbed, and will be able to generate an image from this. Denser parts of the body, like bone, show up as white, as few x-rays could pass through. Sometimes a contrast agent is given to the patient in order for soft tissues to be seen more easily on the image. The x-ray is very quick, and the whole procedure should only take a few minutes.

Advantages

- The machine does not surround the entire body, thus won't cause anxiety in individuals with claustrophobia.
- The procedure solely takes a couple of minutes.

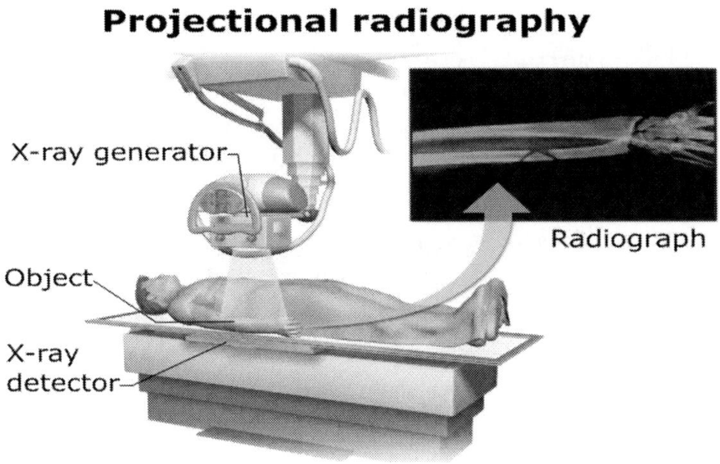

Figure 4. Diagram of how an x-ray scanner works.

Disadvantages

- Some contrast agents could cause unwanted side effects.
- X-rays expose the patient to unwanted radiation, that might doubtless cause cancer, but the amount of radiation given off is marginal.

Applications of MRI Scan in Medical Diagnosis

- Neuroimaging. MRI is the investigative tool of choice for neurological cancers as it is more sensitive than CT for small tumours and offers better visualization of the posterior fossa. The contrast provided between grey and white matter makes it the optimal choice for many conditions of the central nervous system including demyelinating diseases, dementia, cerebrovascular disease, infectious diseases and epilepsy.

- Cardiovascular. Cardiac MRI is complementary to other imaging techniques, such as echocardiography, cardiac CT and nuclear medicine. Its applications include assessment of myocardial ischemia and viability, cardiomyopathies, myocarditis, iron overload, vascular diseases and congenital heart disease.
- Musculoskeletal. Applications in the musculoskeletal system include spinal imaging, assessment of joint disease and soft tissue tumours.
- Oncology. MRI is the investigation of choice in the preoperative staging of rectal and prostate cancer, and has a role in the diagnosis, staging, and follow-up of other tumours.
- Liver and Gastrointestinal MRI. Hepatobiliary MRI is used to detect and characterize lesions of the liver, pancreas and bile ducts. Focal or diffuse disorders of the liver may be evaluated using diffusion-weighted, opposed-phase imaging and dynamic contrast enhancement sequences. Extracellular contrast agents are widely used in liver MRI and newer hepatobiliary contrast agents also provide the opportunity to perform functional biliary imaging. Anatomical imaging of the bile ducts is achieved by using a heavily T2-weighted sequence in magnetic resonance cholangiopancreatography (MRCP).
- MRI uses three electromagnetic fields: a very strong static magnetic field to polarize the hydrogen nuclei, called the static field; a weaker time-varying (on the order of 1 kHz) field for spatial encoding, called the gradient field; and a weak radio-frequency (RF) field for manipulation of the hydrogen nuclei to produce measurable signals, collected through an RF antenna.

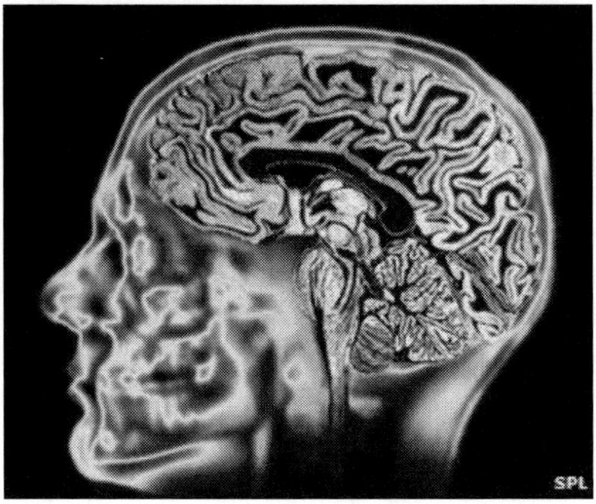

Figure 4. Neuroimaging.

Advanced Technologies for Medical Imaging Systems

- Compute options. OEMs and ODMs are designing solutions across a diverse range of form factors, from mobile ultrasound to nuclear medicine. Each form factor has its own unique compute requirements. Intel offers a range of CPUs that meet the needs of any imaging modality.
- Scalable platforms. The medical imaging industry is moving toward more standardized computing platforms that can be shared across modalities to lower costs and accelerate innovation. Intel supports scalability with an unmatched product portfolio that includes compute, storage, memory, and networking, backed by extensive software resources. We can help with resources for full-stack platforms that support today's most challenging workloads and tomorrow's designs.
- Edge-to-cloud deployment. The ability to flexibly deploy workloads at the edge and in the cloud is quickly becoming the status quo. Edge computing in particular allows providers to analyse images and make decisions at the point of care. This near-real-time medical image analysis can enable faster decisions that help save lives. With our broad range of solutions

and deep partner ecosystem, Intel enables a modern edge-to-cloud architecture.

- Artificial intelligence**.** AI is revealing new ways to streamline workflows and is becoming a de facto capability in imaging systems. For example, deep learning in medical imaging can help prioritize images for a patient with a potentially fatal brain bleed over others in the queue. In other cases, AI can help evaluate images quickly and accurately while removing variances. AI can even help with patient positioning, which can mean the difference between a useful image and the inconvenience of a retake. Intel offers hardware with built-in AI acceleration, plus optimized software. Our demonstrated success in AI ranges from helping deploy deep learning models on in-market modalities to designing mobile systems that can run AI without connectivity.
- Developer enablement. To gain traction in the medical imaging marketplace, software developers building AI-enabled applications need to provide complete solutions and differentiated capabilities. Applications must meet an extensive range of accuracy, cost, customer deployment, and scalability requirements. Intel offers a broad array of tools and programs designed to help software developers successfully develop, launch, and scale new medical imaging AI applications that meet the needs of the healthcare market.

Medical Imaging in the Cloud

There has been growing trend to migrate from on-premise PACS to a Cloud Based PACS. A recent article by Applied Radiology said, "As the digital-imaging realm is embraced across the healthcare enterprise, the swift transition from terabytes to petabytes of data has put radiology on the brink of information overload. Cloud computing offers the imaging department of the future the tools to manage data much more intelligently."

The Future of Medical Imaging: AI, Cloud and Beyond

While medical sensors have played a key role in imaging in the last two decades, future approaches will revolve around computation and more-intensive compute power. Computation and AI make image gathering more efficient and shorten image acquisition times. In addition, the field will likely see more cloud-hosted medical imaging data.

"We're seeing a lot more movement toward cloud-hosted applications and technology using computer servers and more-intensive compute power that can be hosted remotely," Harris says, adding that he would like to see AI algorithms cut down imagine processing time as well. "We're looking at trying to replace some of that time-intensive work with a better segmentation that will allow us to produce the results in less time," Harris says. "We have some cases that take us one to two hours, and if we can cut that time in half using advanced algorithms, that would be really great."

In addition, AI will help radiologist's spot images they would not be able to see with the human eye. "There's a tremendous amount of data in the images that is currently lost because the human eye can't process it," Harris says. "With the help of AI, there's a tremendous amount of information that could be gleaned quantitatively from that data and presented to the radiologist and referring physicians to help with diagnosis and treatment planning."

With technologies like 3D medical imaging and artificial intelligence at doctors' disposal, medical professionals can become "superhuman" and 3D imaging is really "incredible superpowers."

Conclusion

New CT and magnetic resonance imaging systems will scale the image resolution and also the reconstruction time. Whole human body scans with this resolution reach many Gigabytes of data load. Large medical image data occurs in two different ways: first, a large quantity of image data from thousands of pictures like in image archiving and communication systems and second, and an outsized quantity of image data from a single data set.

Medical image computing generally operates on uniformly sampled data with regular 2D and volumes in 3D, generically referred to as pictures. At every sample point, data is commonly represented in integral form such as signed and unsigned short (16-bit), although forms from unsigned char 8-bit to 32-bit float are not uncommon. The particular meaning of the data at the sample point depends on modality: for example a CT acquisition collects radio density values, while a MRI acquisition cloud collect T1 or T2-weighted images. Longitudinal, time-varying acquisitions may or may not acquire images with regular time steps. Fan-like images due to modalities such as curved-array ultrasound are also common and require different representational and algorithmic techniques to process. Other data forms include sheared images due to gantry tilt during acquisition; and unstructured meshes, like hexahedral and tetrahedral forms, that are utilized in advanced biomechanical analysis. Segmentation is the process of partitioning a picture into different segments. In medical imaging, these segments often correspond to different tissue classes, organs, pathologies, or other biologically relevant structures. Medical image segmentation is made difficult by low contrast, noise, and different imaging ambiguities. Although there are many computer vision techniques for image segmentation, some have been adapted specifically for medical image computing. Below cloud is a sampling of techniques within this field; the implementation depends on the expertise that clinicians can provide.

The commonly used term "biomedical image processing" means the provision of digital image processing for biomedical sciences. In general, digital image processing covers four major areas

1. Image formation includes all the steps from capturing the image to forming a digital image matrix.
2. Image visualization refers to all or any sort of manipulation of this matrix, leading to an optimized output of the image.

3. Image analysis includes all the steps of processing, which are used for quantitative measurements as well as abstract interpretations of biomedical images. These steps require a priori knowledge on the nature and content of the images, which must be integrated into the algorithms on a high level of abstraction. Thus, the process of image analysis is very specific, and developed algorithms can be transferred rarely directly into other application domains.
4. Image management sums up all techniques that provide the efficient storage, communication, transmission, archiving, and access (retrieval) of image data. Thus, the methods of telemedicine are also a part of the image management. In contrast to image analysis, which is usually also mentioned as high-level image processing, low level processing denotes manual or automatic techniques, which may be realized without a priori knowledge on the specific content of images. This sort of algorithms has similar effects no matter of the content of the images. For instance, histogram stretching of a radiograph improves the contrast because it does on any holiday photograph. Therefore, low-level processing methods are usually available with programs for image enhancement.

Algorithms used for Image Processing at Different Stages Image registration Image registration is one among the foremost common algorithms in medical imaging and therefore one with the foremost GPU implementations. One reason for this is often the GPU's hardware support for linear interpolation, which makes it possible to rework images and volumes very efficiently. Hastreiter and Ertl (1998) were one among the primary to require advantage of the GPU for image registration, mainly for its ability to perform fast interpolation in 3D. A standard approach is to let the GPU calculate a similarity measure, often mutual information over the pictures in parallel while the CPU runs a serial optimization algorithm to seek out the parameters that give the simplest match between the two images.

References

Abutalebi J, Rosa PA, Tettamanti M, Green DW, Cappa SF. Bilingual aphasia and language control: a follow-up fMRI and intrinsic connectivity study. *Brain Lang.* 2009; 109:141–156.

American College of Radiology; American Society of Neuroradiology (2010), "ACRASNR practice guideline for the performance of computed tomography (CT) of the brain," *Agency for Healthcare Research and Quality*, Reston, VA, USA: American College of Radiology, archived from the original on 15 September 2012, retrieved 9 September 2012.

Axioms and fundamental equations of image processing, *Arch. Rat. Mech. Ana*l. 123(1993), no.3, 199–257.

Babiloni C, Pizzella V, Gratta CD, Ferretti A, Romani GL. Fundamentals of electroencephalography, magneto-encefalography, and functional magnetic resonance imaging. *Int Rev Neurobiol*. 2009; 86:67–80.

Banerjee R, Pavlides M, Tunnicliffe EM, Piechnik SK, Sarania N, Philips R, Collier JD, Booth JC, Schneider JE, Wang LM, Delaney DW, Fleming KA, Robson MD, Barnes E, Neubauer S (January 2014). "Multiparametric magnetic resonance for the non-invasive diagnosis of liver disease". *Journal of Hepatology*. 60 (1): 69–77. doi:10.1016/j.jhep.2013.09.002. PMC 3865797. PMID 24036007.

Barker PB, Soher BJ, Blackband SJ, Chatham JC, Mathews VP, Bryan RN. Quantitation of proton NMR spectra of the human brain using tissue water as an internal concentration reference. *NMR Biomed.* 1993; 6:89–94.

Brenner DJ, Hall EJ (November 2007). "Computed tomography--an increasing source of radiation exposure". *The New England Journal of Medicine.* 357 (22): 2277–84. doi:10.1056/NEJMra072149. PMID 18046031.

Brown GG, Clark C, Liu TT. Measurement of cerebral perfusion with arterial spin labeling: Part2. Applications. *Journal of the International Neuropsychological Society*. 2007; 13:526–538.

Brulatout S, Meric P, Loubinoux I, Borredon J, Correze JL, Roucher P, Gillet B, Berenger G, Beloeil JC, Tiffon B, Mispelter J, Seylaz J. A one-dimensional (proton and phosphorus) and two-dimensional (proton) in vivo NMR spectroscopic study of reversible global cerebral ischemia. *J Neurochem*. 1996; 66:2491–2499.

Buxton RB. *Introduction to functional magnetic resonance imaging: Principles and techniques*. New York: Cambridge University Press; 2001.

Chen R, Cros D, Curra A, Di Lazzaro V, Lefaucheur JP, Magistris MR, Mills K, Rösler KM, Triggs WJ, Ugawa Y, Ziemann U. The clinical diagnostic utility of transcranial magnetic stimulation: report of an IFCN committee. *Clin Neurophysiol*. 2008; 119:504–32.

Comley RA, Kallend D (February 2013). "Imaging in the cardiovascular and metabolic disease area". *Drug Discovery Today*. **18** (3–4): 185–92. doi:10.1016/j.drudis.2012.09.008. PMID 23032726.

De SN, Matthews PM, Arnold DL. Reversible decreases in N-acetylaspartate after acute brain injury. *Magn Reson Med*. 1995; 34:721–727.

de Jongh A, de Munck JC, Baayen JC, Puligheddu M, Jonkman EJ, Stam CJ. Localization of fast MEG waves in patients with brain tumors and epilepsy. *Brain Topogr*. 2003; 15:173–179.

De Luca M, Beckmann CF, De Stefano N, Matthews PM, Smith SM. fMRI resting state networks define distinct modes of long-distance interactions in the human brain. *NeuroImage*. 2006; 29:1359–1367. [PubMed] [Google Scholar].

Eckert MA, Tenforde A, et al. To modulate or not to modulate: differing results in uniquely shaped Williams syndrome brains. *Neuroimage.* 2006; 32:1001–7. [PubMed] [Google Scholar].

Eliassen JC, Boespflug EL, Lamy M, Allendorfer J, Chu WJ, Szaflarski JP. Brain-mapping techniques for evaluating poststroke recovery and rehabilitation: a review. *Top Stroke Rehabil*. 2008; 15:427–450. [PMC free article] [PubMed] [Google Scholar].

Ernst TR, Kreis R, Ross BD. Absolute quantitation of water and metabolites in the human brain. I. Compartments and water. *J Magn Reson*, Series B. 1993; 102:1–8. [Google Scholar].

Filippi MM, Oliveri M, Pasqualetti P, Cicinelli P, Traversa R, Vernieri F, Palmieri MG, Rossini PM. Effects of motor imagery on motor cortical output topography in Parkinson's disease. *Neurology*. 2001; 57:55–61. [PubMed] [Google Scholar].

Foltys H, Krings T, Meister IG, Sparing R, Boroojerdi B, Thron A, Topper R. Motor representation in patients rapidly recovering after stroke: a functional magnetic resonance imaging and transcranial magnetic stimulation study. *Clin Neurophysiol.* 2003; 114:2404–2415. [PubMed] [Google Scholar].

Gerloff C, Bushara K, Sailer A, Wassermann EM, Chen R, Matsuoka T, Waldvogel D, Wittenberg GF, Ishii K, Cohen LG, Hallett M. Multimodal imaging of brain reorganization in motor areas of the contralesional hemisphere of well recovered patients after capsular stroke. *Brain.* 2006;129:791–808. [PubMed] [Google Scholar].

Gillard JH, Waldman A, Barker P. In: *Clinical MR Neuroimaging: Diffusion, Perfusion, and Spectroscopy.* Gillard Jonathan H, Waldman Adam D, Barker Peter B., editors. Cambridge; New York, NY: Cambridge University Press; 2005. [Google Scholar].

Golay X, Hendrikse J, Lim TC. Perfusion imaging using arterial spins labeling. *Top Magn Reson Imaging.* 2004; 15(1):10–27. [PubMed] [Google Scholar].

Hensel S, Rockstroh B, Berg P, Elbert T, Schonle PW. Left-hemispheric abnormal EEG activity in relation to impairment and recovery in aphasic patients. *Psychophysiology.* 2004; 41(3):394–400. [PubMed] [Google Scholar].

Indefrey P, Levelt WJ. The spatial and temporal signatures of word production components. *Cognition.* 2004; 92(1-2):101–144. [PubMed] [Google Scholar].

Johnsrude I, Hauk O. Neuroimaging: Techniques for examining human brain function. In: Braisby N, editor. *Cognitive Psychology: A Methods Companion.* New York: Oxford University Press; 2005. [Google Scholar].

Kamada K, Saguer M, Moller M, Wicklow K, Katenhauser M, Kober H, et al. Functional and metabolic analysis of cerebral ischemia using magnetoencephalography and proton magnetic resonance spectroscopy. *Ann Neurol.* 1997; 42(4):554–563. [PubMed] [Google Scholar].

Lewine JD, Davis JT, Sloan JH, Kodituwakku PW, Orrison WW., Jr Neuromagnetic assessment of pathophysiologic brain activity induced by minor head trauma. *AJNR Am J Neuroradiol.* 1999; 20(5):857–866.

In: Research Challenges in Science ... ISBN: 978-1-68507-008-3
Editors: S. Kannadhasan et al.

Chapter 3

A Review on Data Driven Based Discrimination Techniques between Faults and Power Swing for Hybrid Renewable Energy Systems

Swati S. Pachauri[1,*], Sonal Tripathi[2], Abdur Rahman Khan[3], Bhaumik Joshi[4], Siddarath Vats[4] and Siddhi Vinayak Pandey[4]

[1]Department of Physics, School of Science, Gujarat University, India
[2]Department of Electrical & Electronics, School of Automation, Banasthali Vidyapith, Rajasthan, India
[3]Department of Mathematics, Siddharth University, Basti, Uttar Pradesh, India
[4]Department of Electrical Engineering, Adani Institute of Infrastructure Engineering, Ahmedabad, Gujarat, India

[*] Corresponding Author's E-mail: swatipachauri98@gmail.com.

ABSTRACT

Due to the limitations of non-renewable resources in terms of energy-efficient and sustainability, the demand for renewable-based energy generating stations is increasing drastically. In order to balance the supply-demand chain, the hybrid grid-connected systems are being used. Presently, most of the hybrid grid-connected systems are replacing their conventional (non-renewable) stations with renewable energy-based generating power plants. With this advancement, the discrimination between faults and power swing has always been a challenging task for all grid connected systems. Sometimes, it leads to the false tripping command to distance protection systems utilized within a transmission line. For such associated issues, the accurate discrimination between faults and power swing can avoid the mal-operation of distance protection scheme and let the systems work more efficiently without any intervention. The present paper provides detailed insights about various discrimination techniques utilized in modern grid-connected systems using data-driven techniques, such as - Support Vector Machine (SVM), Artificial Neural Network (ANN), Neuro-Fuzzy techniques, Bayesian Classifier & Adaptive Wavelet technique, Linear Discriminant Analysis (LDS), and transformations approach. Also, the comparative analysis between conventional-based discrimination approach with modern data-driven-based techniques has been explicated along with future works to be carried out for better optimization.

INTRODUCTION

In recent decades, the world has seen a sudden rise in energy contribution from renewables with a drastic increase in the installed capacity of various hybrid energy farms [1]. The motivation for the sudden change is driven by the instantaneous escalation in energy demand and secondly by, the deteriorating condition of climate change [2]. An electric grid system is an enormous network incorporating generation, transmission and distribution of electric energy, this network supplies electricity from the generator to respective loads [3]. The stability of the grid is necessary to maintain equilibrium between starting (Generators) and ending (Loads) points but such huge integration of renewables in electrical grids increases instability concerns of the overall system. Negligence while treating electrical faults and malfunction of the relays due to power swings can result in local contingencies and eventually can lead to change in power angle of other buses, loss of synchronism, cascade tripping of local power system or even local blackouts [4-5]. As mentioned in [6], to furthermore complicate things, there is the rapid adoption of EVs (Electric Vehicles) in developed nations which in turn induces harmonics and lowers the power quality. The real-world negative effect of the wind turbine while interfacing with power grid by destabilizing it, significantly diminishes voltage and frequency of the overall power system [7]. Yusoff Nooraizzat and Ahmad Farid [8] compared the two different scenarios of the simulated IEEE-9 bus system in the presence and absence of PV generators and found out that the former case showed a decrease in active power when it was compared with the latter and also [9] reviews the adverse impacts of commonly used renewables on the power grid.

Sudden connection of heavy load, disconnection of transmission lines and loss of generation lead to power swing in the system. During the power swing, the active and reactive power of a transmission line gets fluctuated consequently with a large disturbance like fault within a generating/ transmission/ receiving end station. If theses fluctuations of active and reactive power will not be blocked expeditiously then it may cause the wrong operation of distance relay [10-11]. Sometimes it generates nonessential tripping signal for transmission line [12]. The system tries to recuperate from this temporary imbalance state caused due to stable power swings, unlike unstable power swings. Power swing causes short-term disturbance in voltage and power level of the system which sometimes mimics the behavior of three-phase faults. The Three-phase faults can be grouped in open circuit and short circuit condition, open circuit faults occur due to the failure of one or more phase conductors while short circuit faults arise when phase meltdown occurs due to electrical contact of phases with each other or with the ground, this results in a sudden decrease in impedance and allowing high current to flow through the system. It can be a single line to ground (L-g), double line to ground (LL-g), three-phase to ground fault (LLL-g), or line to line fault (L-L). During fault, a sudden spike and drop in current and voltage respectively are observed and a tripping signal is issued to the receiving end of the line by a circuit breaker to cut the flow of power, so the sensitive amenities can be saved from damage due to high current. So, power swing can be inferred as a fault by distance relays and they issue trip signal in line [13]. As a remedy, Power swing blocking (PSB) unit is installed in distance relays to prevent such tripping while power swing [10]. Alternatively, there is a chance of fault during power swing which can be symmetrical or asymmetrical in nature [14], at this instance distance relay should disable the function of PSB and allow the trip signal to pass [15, 16].

In FFT analysis, a parameter is selected which changes during fault and is observed in every cycle. Any deviation from measured value having a difference above the particular pre-decided threshold value is considered as a decisive parameter for fault during power swing [17]. Mahamedi and Guo Zhu [18] used control algorithm for symmetrical fault. For FFT analysis, the frequency of three phase active power has been observed of 50 Hz in real time domain. FFT studies variation in the frequency domain of a signal while WT gives freedom to study both the time and frequency domain of a signal. Brahma in [19] used wavelets to identify power swing and symmetrical fault and transient in the power of the signal. With a sampling frequency of 40.96 kHz, this method gives satisfactory results [18] by Pang and Kezunovic studied the energy of high-frequency components in forward and backward traveling waves generated due to transients in power swing signal i.e., fault. This is a high-speed method even effective at a low sampling rate of 10 kHz, higher sampling rate will give better results. In [21] Mahamedi proposed a method where WT is applied on the magnitude of impedance. With a sampling rate of 1 kHz delay of 30ms is recorded. This method is effective for high voltage and extra-high voltage. Dubey, Samantaray in [22] focused on wavelet energy for the uncovering of symmetric faults during power swing. Lotfifard, Faiz & Kezunovic in [23] considered decaying dc components in current form during faults using the prony model. The [24] discusses a signal processing technique that can distinguish between three faults and power swing using zero frequency filtering of the current waveform, first the current signal was subjected to Hilbert's envelope. This technique comes with the added advantage of identifying faults during power swings [25] by J.

Ganeswara Rao and Ashok Kumar Pradhan describes an auto-regressive technique which is based on differential power-based technique, the described technique was able to discriminate and detect three-phase symmetrical faults precisely irrespective of nature of the power swings [26] is based on a novel approach to detect all types of faults in the series compensated lines inclusive of 9-bus 3-machine powers system which is a negative sequence current based method [27] by Jalal Khodaparast and Mojtaba Khederzadeh confers a de-blocking scheme of the PSB at time of faults during power swing by calculating the transient motor (TM) index and then followed by implementing Fourier-Taylor transform, which will evaluate the dynamic phasor of given signal.

This manuscript comprises of 7 different sections and their description is as follows: Section II gives a detailed description of SVM (Support Vector Machine) technique along with its possible implementation and then this section is followed by three similar successive sections, they are Decision Tree (Section III), Random Forest (Section IV), and Convolutional Neural Network (Section V). Section VI and Section VII describes the possible future work and conclusion of aforementioned review work respectively.

SUPPORT VECTOR MACHINE

Support Vector Machines (SVM) is a supervised learning algorithm which solves binary classification problems by formulating them as convex optimization problems. The optimization problem is to find the maximum optimum margin separating the hyperplanes, while truly predicting as many training points as possible. The optimal hyperplane is represented by support vectors. An SVM classifier makes the greatest margin hyperplane that lies in changing input space and parts the model classes while amplifying the distance to the closest neatly split models as shown in Figure 1. The boundaries of the arrangement hyperplane are gotten from a quadratic programming advancement issue [28]. Here preparing vectors are planned into a higher (possibly limitless) dimensional space by a capacity. SVM tracks down a straight isolating hyperplane with the maximal margin in this higher-dimensional space. It uses a loss function for error reduction, and "Kernel" is utilized because a bunch of numerical capacities utilized in Support Vector Machine gives the window to control the information. Kernels can be classified as follows: linear, polynomial, radial basis function, and sigmoid [29]. Since SVMs depend on the structural risk minimization (SRM) guideline, it offers a bigger speculation capacity contrasted with the other managed learning strategies like the artificial neural network (ANN) [30].

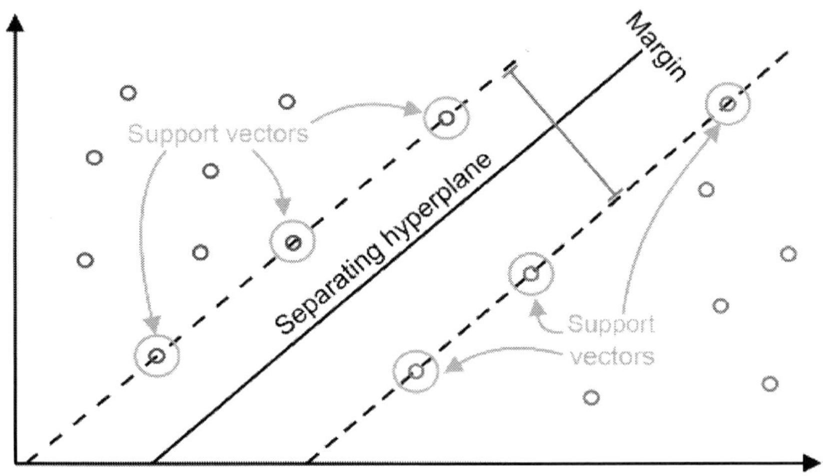

Figure 1. Support Vector Machines.

Two classifiers, taking into account SVM, have been utilized. SVM-1 is used for speedy discovery of the faults occurring in the power system and SVM-2 is utilized for perceiving the power swings and voltage shakiness events. SVM-1 uses the possibility of the pace of progress of moving sums for isolating a fault condition from various events, for instance, power swings, or voltage unsteadiness. Neighborhood assessments are utilized to avoid time delays in the action of the exchange. Various information sources are picked to ensure the correct action of the arrangement like positive-arrangement, receptive power, voltage and current phasors. The speed of progress in these sums are taken as the contributions to the SVM-1 and it is handled by differentiating the phasor values and that of the past cycle. The contributions to the SVM-2 are gotten from locally open information, close by the PMU information gained from the less than desirable finish of the transmission line. The PMU phasor rate is picked as 60 phasors/s for the 60-Hz structure. The distinctive assessment sums open at the exchange territory, the greatness of the voltage at the hand-off area, the size of the current through the line, genuine power stream, and responsive power misfortunes in the line are taken as contributions to the classifier. Responsive power misfortune is handled from the PMU assessments at the area and the far-off completion of the transmission line. Ten cycles' information has been taken to shape the component vector for the SVM-2, including highlights [30] Nonetheless, because of the essential of voltage and current signals, this arrangement ends up being seriously astounding. In addition, it is endorsed with confined trials without contemplating sensible circumstances (varying fault, system boundaries, and subtle dataset).

Any aggravation in a framework can originate a disparity linking the mechanical power contribution to the turbine and the electrical power yield of the generator. The administering system of the machine can attempt to adapt to this lopsidedness however it will take a specific unmistakable measure of time during which the system will notice oscillations. The possibility of the proposed calculation has been tried on a current piece of a 220 kV Indian power system network model utilizing the PSCAD/EMTDC software bundle, whereby various faults were created in the line and different parameters were recorded, simulation of power swing scenarios were also carried out. [31, 32]. At first, SVM is trained disconnected utilizing a preparation dataset comprising of 5750 simulation cases. Subsequent to arranging SVM, the anomaly recognition algorithm separates between strange conditions and the typical working state. At whatever point a fault happens, samples of post-fault signals are given in equal (at the same time) to two unique algorithms in particular 1) power swing/fault segregation 2) distance estimation algorithm. Here, the distance estimation algorithm requires (N + 2) samples, where N is the quantity of samples in a single cycle. The proposed power swing/fault separation algorithm requires a large portion of a cycle to segregate fault or power swing. In the most noticeably awful situation, if power swing is distinguished by the proposed conspire during which the distance assurance algorithm wrongly sees a first zone fault, the proposed plot needs to give a hindering order to the distance estimation algorithm to stay away from the bogus outing. Simultaneously, the algorithm gets back to the information procurement framework to keep checking for the following inspecting window. This requires the issue of obstructing order by the proposed algorithm preceding the choice of first zone fault (N + 2 samples time) as given by the distance estimation algorithm [33]. In general, a general segregation exactness of 98.71% exhibits the viability of the proposed conspire in practically all functional situations as talked about above. The proposed plot gives a segregation accuracy of over 98% for all issue cases. Besides, it additionally gives better dependability (block hand-off activity) on account of force swing giving over 98% segregation accuracy. The outcome shows that the proposed SVM-based strategy is exceptionally precise to separate deficiencies during power swing conditions which are hard to identify by the ordinary plans dependent on WT and artificial intelligence procedures [33].

Decision Tree

As of late, DT has arisen as a piece of mainstream data-mining approach for various classification assignments in light of its attributes identified with simpler execution and improved execution on bigger datasets. Be that as it may, the forecast exactness of DT is regularly uncompetitive when contrasted with other information mining strategies like SVM and deep neural networks. Frequently for complex datasets, DT doesn't give good execution during testing. This issue is settled by using the troupe of the tree rather than a single DT. The collection of the individual outputs of the DTs in the gathering to show up at the ultimate result defeats the limit identified with conceivable biasing by a single DT. The most well broadly utilized gathering procedure is bootstrap total generally alluded to as bagging, in which the quantity of subsets is haphazardly drawn by dividing the whole training information by utilizing examining with substitution. Training of subsets is done by a similar student and comparing to every individual subset, a different classifier is prepared. Bagging joins the mix of results made by singular DTs by following the surveying system and the last output is gotten by the dominant part deciding in favor of classification and averaging for regression. The use of BT for flaw identification, classification, and the area has been depicted in the following sub-segments [34, 35].

The creation of a decision tree incorporates building up a tree from learning tests until terminal hubs contain the instances of only one class. An order tree is built ward on the parting rule. Various parting models are used for arrangement from which the Gini parting guideline is extensively used. Gini parting standard uses a limit called contamination work implied by I(n) for hub n. Toxin at hub n is yielded (1).

$$I(n) = -\sum_{i=1}^{k} p(wi|n) \log p(wi|n) \tag{1}$$

where p($wi|n$) is the extent of the example assigned to class wi at node n. The debasement work in the Gini splitting standard is given by (2)

$$I(n) = \sum p(k|n)\, p(l|n) \tag{2}$$

where k, l = 1, ..., k is the file of the class, and p(kln) is the restrictive likelihood of class k at node n. Gini splitting standard gets the biggest class from all the preparation tests and segregates it from the remainder of the examples. In the wake of making the decision tree, the following assignment is to choose the right tree for classification [36, 37]. The below figure is a visual representation of a decision tree.

The current signs in both the circuits are pre-arranged through the most un-square and Adaline calculation for the extraction of the DC balance boundaries. A moving window of one cycle is considered for constant evaluation of DC counterbalance size and time consistency. The DC balance boundaries and essential piece of post-issue current signs contain the part vector for setting up the packing tree (BT)- based shortcoming pointer/classifier, critic, broken zone identifier (FZI), and force swing finder. BT contains a lot of DTs, wherein the portrayal and gauge task is developed through finishing a popularity-based method on the yields of the individual DTs.

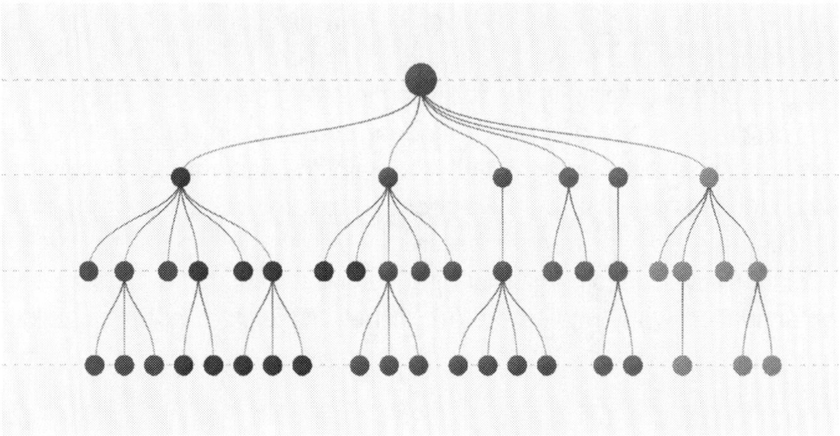

Figure 2. Decision Tree.

For the FD/C task, each classifier module arranged for each issue type goes through equivalent taking care of to predict such a deficiency. BT's, BTZ1 to BTZ5, address the zone identifier module for the LG, twofold 2LG (LG), 2L, 3LG, 3L, independently, for circuit 1. Moreover, for circuit 2, the zone identifier modules contrasting with LG, 2LG, 3LG, 2L, 3L are tended to by BTs, BTZ6 to BTZ10, independently. At the point when the deficiency is recognized and the broken zone has been perceived, a staggering sign is made while because of no-issue, the force swing pointer comes directly into it. The revelation of force swing is immediately followed by the time of discouraging sign to the exchange to avoid undesired staggering. The exhibition examination of the created conspire has been suitably done on a 400 kV three-stage twofold circuit line and WSCC 9-transport system. The presentation appraisal reveals that the created conspire is prepared for playing out the task of FD/C, flawed zone recognizing evidence, and territory evaluation constantly and decisively inside one cycle, for various causes related to the assortment in shortcoming and force swing circumstances. The proposed conspire has similarly been endorsed on a consistent stage using the OPAL-RT progressed test framework to verify its sufficiency for field applications. [35].

Crucial parts of voltage and current are taken as contributions to the DT for power swing and fault location. For fault order, key parts of zero sequence parts of current are taken as information. The benefit of the DT is in fact it isn't influenced by the length of the dataset utilized for preparing. The unwavering quality of this plan isn't influenced by various fault conditions like variety in fault distance, fault type, fault resistance, fault inception angle, and so on the results exhibit that the suggested conspire viably identifies the fault type and the fault inside half-cycle time. The precision of the suggested strategy is 99.99% for all fault cases [37].

Random Forest

The random forest algorithm, proposed by L. Breiman in 2001, has been amazingly effective as a universally useful classification and regression technique. The methodology, which consolidates a few randomized choice trees and totals their expectations by averaging, has shown incredible execution in settings where the quantity of factors is a lot bigger than the number of perceptions. In addition, it is sufficiently flexible to be applied to huge scope issues, is effortlessly adjusted to different specially appointed learning assignments, and returns proportions of variable significance. Accentuation is set on the numerical powers driving the algorithm, with unique consideration given to the choice of boundaries, the resampling component, and variable significance measures. What has incredibly added to the prevalence of forests is the way that they can be applied to a wide scope of forecast issues and have not many boundaries to tune. Besides being easy to utilize, the technique is by and large perceived for its precision and its capacity to manage little example sizes and high-dimensional element spaces. Simultaneously, it is effectively parallelizable and has along these lines the possibility to manage enormous genuine frameworks [38].

Random Forest is an ensemble learning algorithm. RF amasses various decision trees and sums them together to get more strong and exact estimates. It relies upon two stunning AI techniques, explicitly, bagging and random element choice [39, 40]. In bagging, different preparing sets are randomly evolved from the dataset using exchange and for every arrangement set, another decision tree is readied. The complete technique for larger part projecting a voting form is used for the definitive gauge. Moreover, random forest decorrelates the individual trees by considering a random illustration of m markers which are picked as a split contender from the full course of action of p pointers. The random assurance of m pointers is reiterated at each split of the decision tree [41].

Principally, make a forest of decision trees, whereby every decision tree is made by a free information point from the preparation set. For each hub select m factors randomly from every conceivable indicator, later track down the ideal split among these m factors. At that point subsequent to developing the tree to greatest profundity. Utilize prevailing part projecting a polling form of the trees ordinary to get forecasts for new data. The RF predicts execution during the planning steps itself in regards to the OOB (Out of Bag) check of mistake rate. During each bootstrap cycle, expect execution with the data not used to fit a given stowed tree (implied as out-of-pack data). OOB forecasts' complete gives the OOB check of blunder rate. RF is fiery, less computationally capricious, diminishes the tendency and contrast due to bootstrap mixture, and doesn't tend to overfit [41].

To prepare and approve the RF model, 9216 distinctive fault cases were copied using PSCAD/EMTDC programming. These fault sets were reenacted for unsymmetrical (A-g, AB, BC-g) and adjusted (ABC-g) fault types at various swing frequencies, starting focuses at various fault regions, and for a couple of fault insurances. The continuous measure of the current sign is taken out from one completion of the transmission line for extra examination. RF fault order model is prepared with the absolute differentiation current of each stage as information features and fault type as the yield class. A total of 27648 models covering 9216 fault cases is used for model preparation. Taking into account the out-of-sack mistake rate change governs, an ideal number of trees required for the RF model is recognized as 250. k-Fold cross-endorsement close by the lion's offer projecting a polling form system is used to approve the show of the fitted RF classifier [42].

One finish of the transmission line slow measure of current is used as the RF model's data incorporates. The steadfastness of the RF-based fault order model is attempted and prepared for the change in fault type, swing recurrence, the beginning point, fault obstruction, and fault distance. The results show doubtlessly that, fault order accuracy using the RF computation is 99.8%. The producer's accuracy and customer's precision results exhibit sublime fault type-wise estimate exactness and trustworthiness of arrangement. Moreover, fault grouping is cultivated in under 10ms. It is surmised that the proposed approach reliant upon the random forest to portray the faults during the power swing is fast and careful.

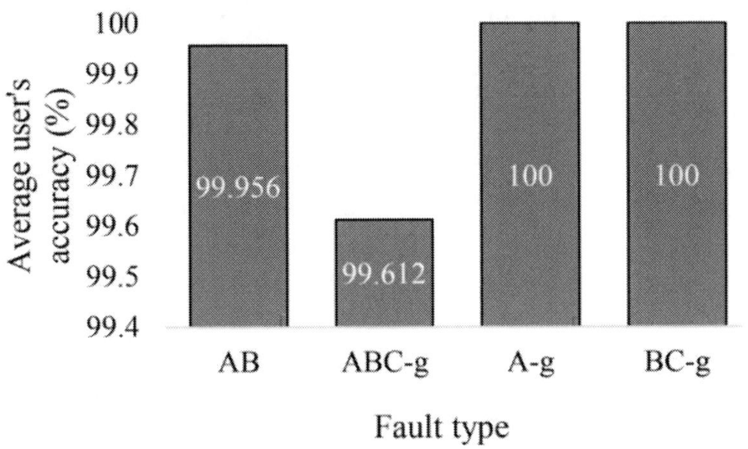

Figure 3. Accuracy of Random Forest technique in different Fault type.

Convolutional Neural Network

The accurate measurement of SOC is a difficult task in real time domain. Because the straightforward measurement method is not applicable for estimation of SOC in live condition. Kalman Filter can be used to overcome this challenge [6]. As mentioned in Zhang 2011 [7] and He 2013 [8] - with the combination of Coulomb Counting technique, Kalman filter estimates the accurate overvoltage and current in real time domain.

Convolutional Neural Network has had momentous outcomes over the previous decade in an assortment of fields identified with pattern recognition; from image processing to voice recognition. The most helpful part of CNN's is decreasing the number of boundaries in ANN. This accomplishment has incited the two scientists and designers to move toward bigger models to tackle complex undertakings, which was unrealistic with exemplary ANNs; The main presumption about issues that are settled by CNN ought not to have highlights that are spatially reliant. All in all, for instance, in a face recognition application, we don't have to focus on where the faces are situated in the images. The solitary concern is to distinguish them regardless of their position in the given images. Another significant part of CNN is to get theoretical highlights when input proliferates toward the more profound layers. For instance, in image arrangement, the edge may be recognized in the principal layers, and afterward the more straightforward shapes in the subsequent layers, and afterward, the more elevated level highlights like faces in the following layers [43, 44]. The main layer in CNN is the convolution layer which takes more often than not inside the network. Network execution likewise relies upon the number of levels inside the network. Yet, then again as the quantity of levels expands the time needed to prepare and test the network. Today the CNN considers a force full apparatus inside AI for a ton of utilization, for example, face identification and image, video recognition, and voice recognition [45]. Convolution is the basic utilization of a channel to an input that outcomes in enactment. Rehashed utilization of a similar channel to an input brings about a guide of initiations called a component map, demonstrating the areas and strength of an identified element in input, like an image. The fundamental thought of pooling is down-testing to decrease the intricacy for additional layers. Max-pooling is perhaps the most well-known sorts of pooling strategies. It segments the picture into sub-locale square shapes, and it just returns the maximum worth within that sub-district. In addition, pooling can be utilized with non-equivalent filters and strides to improve effectiveness. For instance, a 3x3 max-pooling with stride 2 keeps a few covers between the spaces [45, 46]. Convolutional layer, followed by Max-polling layer, followed by n Dense layers, and an output layer, is the common architecture of a Convolutional Neural Network. Figure 3 is the model representation of a Convolutional Neural Network [47, 48].

The accurate measurement of SOC is a difficult task in real time domain. Because the straightforward measurement method is not applicable for estimation of SOC in live condition. Kalman Filter can be used to overcome this challenge [6]. As mentioned in Zhang 2011 [7] and He 2013 [8] - with the combination of Coulomb Counting technique, Kalman filter estimates the accurate overvoltage and current in real time domain.

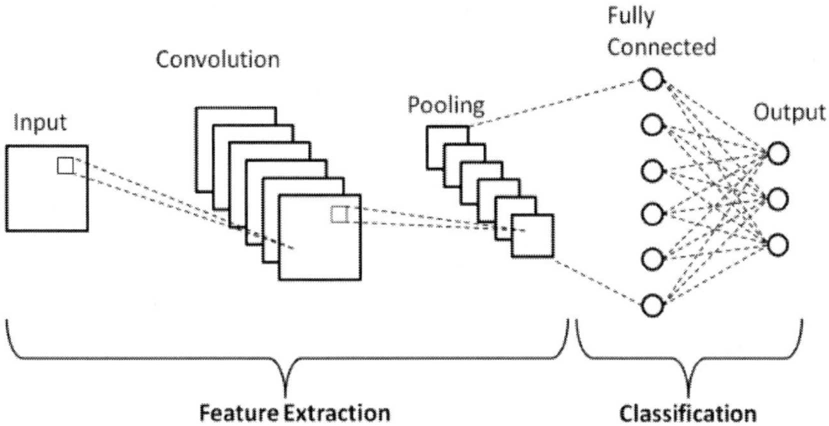

Figure 4. Convolutional Neural Network.

The dataset gathered is pre-processed utilizing the Probabilistic Principal Component Analysis (PPCA) strategy joined with the Partial Least Squares Regression (PLSR) calculation, for information decrease where the size of the information is diminished. In the PPCA model, each piece of information is expressed with probabilistic appropriation, and another model is recreated trailed by the straight blend between the primary part and the information. In the wake of pre-processing, the qualities separated from the recreation are named Normal, Fault, Power Swing and are given for preparing to the CNN. The prepared CNN classifier at that point is called to order the system condition took care of with the test tests. The proposed system is demonstrated and recreated with MATLAB/SIMULINK. The SIMULINK model is introduced in the displaying of the IEEE 9-bus system. All the necessary information achieved from the IEEE 9-bus system which uses PMUs are prepared after the pre-processing stage to be given to the CNN classifier. The execution and preparation of the CNN. The engineering incorporates an input with 6 information esteems from the output of the pre-processing for each case. There is a sum of 78 perceptions/cases as referenced previously. Then, this input is increased by an arbitrarily created weight [49, 50]. The increased input is added with a predisposition for biasing the initiation work. There are 10 hidden hubs utilized in the hidden layer. From the hidden layer, the qualities are then passed to the output layer for classification. In the output layer, the arbitrary weight is increased and inclination is included in a similar way. The output layer comprises three hubs (class labels) for Normal, Fault, and Power Swing conditions [51]. The proposed CNN classifier model is additionally approved in the TensorFlow framework for the equivalent preprocessed dataset. TensorFlow CNN framework is created in the Python Jupiter climate. The exactness accomplished for classification utilizing the TensorFlow and the MATLAB CNN models is 98.7179% and 98.72% individually. Hence, can be presumed that the proposed model of the CNN classifier is ended up being better contrasted with other information-digging models for the tended to issue. The proposed CNN-based classification method gave precise outcomes for all tried conditions without trading off exactness.

Future Works

The expansion of the power system brings new challenges and complications in the system which in turn will cause an increased need for more sophisticated machine learning algorithms to discriminate between three-phase faults, power swings, and other electrical phenomena. With this regard, the following points describe the additional appendage which can be done to the current work.

1. The ANNs (Artificial Neural Network) technique can be employed because the neurons can be trained with high error tolerance and parallel processing capability.
2. Another algorithm that can look upon is LSTM (Long short-term memory), which is a deep learning technique whose architecture is RNN (Recurrent Neural Network) in nature. The advantage of this method is observed when the individual tries to classify the time series data.
3. KNN (K Nearest Neighbor) is one of the simplest classification algorithms that have non-parametric characteristics. It can be

applied where the dataset is simple and can easily be classified.

4. The next classifier is Naive Bayes which belongs to the family of "probabilistic classifiers" and their implementation is based on the Bayes Theorem. It is usually applied when the nature of data points is unknown (Whether it is in continuous or discrete form).

Conclusion

This chapter reviews various schemes and techniques which were built on top of these four data-driven approaches, Support Vector Machines (SVM), Decision Tree (DT), Random Forest (RF), and Convolutional Neural Networks (CNN), for classification of fault and power swing in a power system. The highest accuracy was seen by [iv] decision tree, with an accuracy score of 99.99% for all the tried fault cases. The fastest, as well as precise determination of fault, was achieved using the Random Forest algorithm, with an accuracy score as high as 99.8% with fault classification is accomplished in under 10ms. SVM also can accurately determine a power swing scenario in an interconnected power swing, with an accuracy score of 98% approx. The fourth and last Convolutional Neural Network takes significant training time compared to other algorithms due to its complex structure But, it performs well during the classification of faults with 98.7% of accuracy score on simultaneous testing of all three different cases i.e., normal scenario, fault scenario and power swing scenario in real time domain.

References

[1] Siddhi Vinayak, Pandey. Shriharshini, Ganapuram and Alok Kumar, Singh. 2021. "A Review on Advancements in Photovoltaic Cells." *Advances in Interdisciplinary Engineering: Select Proceedings of FLAME 2020:* 227-236.

[2] Siddhi Vinayak, Pandey. Jeet, Patel and Jay, Patel. 2021. "Battery Storage System in Automobiles: A Review." *Advances in Interdisciplinary Engineering: Select Proceedings of FLAME 2020:* 573-581.

[3] Salmeron, Javier. Kevin, Wood and Ross, Baldick. 2004. "Analysis of electric grid security under terrorist threat." *IEEE Transactions on power systems* 19(2): 905-912.

[4] Wang, Liancheng and Adly A. Girgis. 1997. "A new method for power system transient instability detection." *IEEE Transactions on Power Delivery* 12(3): 1082-1089.

[5] Mechraoui, A., and Thomas, D. W. P. 1995. "A new blocking principle with phase and earth fault detection during fast power swings for distance protection." *IEEE transactions on power delivery* 10(3): 1242-1248.

[6] Blackburn, J. L. 1987. *Protective Relaying: Principles and Applications,* Mercel Dekker. Inc., New York.

[7] Schmietendorf, Katrin. Joachim, Peinke and Oliver, Kamps. 2017. "The impact of turbulent renewable energy production on power grid stability and quality." *The European Physical Journal* B 90(16): 1-6.

[8] Yusoff, Nooraizzat and Ahmad Farid, Abidin. 2013. "The effect of solar Photovoltaic (PV) system to the characteristic of power swing." *In 2013 IEEE 3rd International Conference on System Engineering and Technology*: 99-102.

[9] Asija, Divya. Pallavi, Choudekar and Malvika, Chouhan. 2015. "Performance evaluation and improvement in transient instability of IEEE 9 bus System using exciter and governor control." *Procedia Computer Science* 70: 733-739.

[10] Bakar, Ab Halim Abu. Fazilah Mat, Yatim. Sallehuddin, Yusof and Mohd, Ridzal Othman. 2010. "Analysis of overload conditions in distance relay under severe system contingencies." *International Journal of Electrical Power & Energy Systems* 32(5): 345-350.

[11] Thakallapelli, Abilash. Rachit, Mehra and H. A., Mangalvedekar. 2013. "Differentiation of faults from power swings and detection of high impedance faults by distance relays." In *2013 IEEE 1st International Conference on Condition Assessment Techniques in Electrical Systems (CATCON)*, pp. 374-377. IEEE.

[12] Chothani, Nilesh. Bhavesh, Bhalja and Urmil, Parikh. 2014. "New support vector machine-based digital relaying scheme for discrimination between power swing and fault." *IET Generation, Transmission & Distribution* 8(1): 17-25.

[13] Zhang, Nan and Mladen, Kezunovic. 2005. "A study of synchronized sampling-based fault location algorithm performance under power swing and out-of-step conditions." *In 2005 IEEE Russia Power Tech,* pp. 1-7. IEEE.

[14] AREVA. *MiCOMho P443 fast multifunction distance protection.* (2006).

[15] A, Mechraoui and Thomas, D. W. P. 1997. "A new principle for high resistance earth fault detection during fast power swings for distance protection." *IEEE Transactions on Power Delivery* 12(4): 1452-1457.

[16] Lin, Xiangning. Yan, Gao and Pei, Liu. 2007. "A novel scheme to identify symmetrical faults occurring during power swings." *IEEE Transactions on Power Delivery* 23(1): 73-78.

[17] Karegar, H. Kazemi and Mohamedi, B. 2009. "A new method for fault detection during power swing in distance protection." In *2009 6th International Conference on Electrical Engineering/ Electronics, Computer, Telecommunications and Information Technology,* vol. 1, pp. 230-233. IEEE.

[18] Mahamedi, Behnam. 2010. "A very fast unblocking scheme for distance protection to detect symmetrical faults during power swings." In *2010 Conference Proceedings IPEC*, pp. 378-383. IEEE.

[19] Brahma, S. M. 2006. "Use of wavelets for out of step blocking function of distance relays." In *2006 IEEE Power Engineering Society General Meeting*, pp. 5-pp. IEEE.

[20] Pang, Chengzong and Mladen, Kezunovic. 2010. "Fast distance relay scheme for detecting symmetrical fault during power swing." *IEEE Transactions on Power Delivery* 25(4): 2205-2212.

[21] Mahamedi, Behnam. 2011. "A new power swing blocking function based on wavelet transform." In *2011 2nd International Conference on Electric Power and Energy Conversion Systems (EPECS)*, pp. 1-6. IEEE.

[22] Dubey, Rahul. S. R., Samantaray. Ashok, Tripathy. Chitti, Babu and Muhammad, Ehtesham. 2012. "Wavelet based energy function for symmetrical fault detection during power swing." In *2012 Students Conference on Engineering and Systems*, pp. 1-6. IEEE.

[23] Lotfifard, Saeed. Jawad, Faiz and Mladen, Kezunovic. 2009. "Detection of symmetrical faults by distance relays during power swings." *IEEE transactions on power delivery* 25(1): 81-87.

[24] MS, Prabhu. Paresh Kumar, Nayak and Gayadhar, Pradhan. 2019. "Detection of three-phase fault during power swing using zero frequency filtering." *International Transactions on Electrical Energy Systems* 29(1): e2700.

[25] Rao, J. Ganeswara and Ashok Kumar, Pradhan. 2012. "Differential power-based symmetrical fault detection during power swing." *IEEE transactions on power delivery* 27(3): 1557-1564.

[26] Nayak, Paresh Kumar. Ashok Kumar, Pradhan and Prabodh, Bajpai. 2013. "A fault detection technique for the series-compensated line during power swing." *IEEE transactions on power delivery* 28(2): 714-722.

[27] Khodaparast, Jalal and Mojtaba, Khederzadeh. 2014."Three-phase fault detection during power swing by transient monitor." *IEEE Transactions on Power Systems* 30(5): 2558-2565.

[28] Armin, Shmilovici. 2005. "Support Vector Machines." *Data Mining and Knowledge Discovery Handbook.* Springer: 257-276.

[29] Hsu, Chih Wei. Chih Chung, Chang and Chih, Jen Lin. 2003. *A practical guide to support vector classification.* 1396-1400.

[30] K, Seethalekshmi. Singh, K. S. N. and S. C., Srivastava. 2012."A classification approach using support vector machines to prevent distance relay maloperation under power swing and voltage instability." *IEEE Transactions on Power Delivery* 27(3): 1124-1133.

[31] Wiot, D. 2004. "A new adaptive transient monitoring scheme for detection of power system events." *IEEE Transactions on Power Delivery* 9(1):42–48.

[32] Mohanty, S. R., A. K., Pradhan. A., Routray. 2008. “A cumulative sum-based fault detector for power system relaying application.” *IEEE Transactions on Power Delivery,* 23(1): 79–86.

[33] Chothani, Nilesh. Bhavesh R., Bhalja and Urmil B., Parikh. 2014. "New support vector machine-based digital relaying scheme for discrimination between power swing and fault." *IET Generation, Transmission & Distribution* 8(1): 17-25.

[34] Shukla, Sunil. Ebha, Koley and Subhojit, Ghosh. 2019."DC offset estimation-based fault detection in transmission line during power swing using ensemble of decision tree." *IET Science, Measurement & Technology* 13(2): 212-222.

[35] Polikar, Robi. 2006. "Ensemble based systems in decision making." *IEEE Circuits and systems magazine* 6(3): 21-45.

[36] Tiryaki, Bulent. 2008."Predicting intact rock strength for mechanical excavation using multivariate statistics, artificial neural networks, and regression trees." *Engineering Geology* 99(1-2): 51-60.

[37] Swetapadma, Aleena and Anamika, Yadav. 2016. "Data-mining-based fault during power swing identification in power transmission system." *IET Science, Measurement & Technology* 10(2): 130-139.

[38] Biau, Gérard and Erwan, Scornet. 2016. "A random forest guided tour." *Test* 25(2): 197-227.

[39] Breiman, Leo. 2000. Some infinity theory for predictor ensembles. *Technical Report* 579, Statistics Dept. UCB.

[40] Breiman, Leo. 1996."Bagging predictors." *Machine learning* 24(2): 123-140.

[41] Patil, Dinesh. Naidu, O. D. Preetham, Yalla and S. Hida. 2019. "An ensemble machine learning based fault classification method for faults during power swing." In *2019 IEEE Innovative Smart Grid Technologies-Asia (ISGT Asia)*, pp. 4225-4230. IEEE.

[42] Kohavi, Ron. 1995. "A study of cross-validation and bootstrap for accuracy estimation and model selection." In *Ijcai,* vol. 14(2): 1137-1145.

[43] Abdel-Hamid, Ossama. Li, Deng and Dong, Yu. 2013. "Exploring convolutional neural network structures and optimization techniques for speech recognition." In *Interspeech,* vol. 11:73-85.

[44] Dumoulin, Vincent, and Francesco Visin. 2016. *A guide to convolution arithmetic for deep learning.* arXiv preprint arXiv:1603.07285.

[45] Albawi, Saad. Tareq Abed, Mohammed and Saad, Al-Zawi. 2017."Understanding of a convolutional neural network." In *2017 International Conference on Engineering and Technology (ICET)*, pp. 1-6. Ieee.

[46] Hasan, Ali. P. S. Pouabe, Eboule and Bhekisipho, Twala. 2017. "The use of machine learning techniques to classify power transmission line fault types and locations." In *2017 International Conference on Optimization of Electrical and Electronic Equipment (OPTIM) & 2017 Intl Aegean Conference on Electrical Machines and Power Electronics (ACEMP)*, pp. 221-226. IEEE.

[47] Livani, Hanif and C., Yaman. Evrenosoglu. 2013."A machine learning and wavelet-based fault location method for hybrid transmission lines." *IEEE Transactions on Smart Grid* 5(1): 51-59.

[48] Morales, J. A., E., Orduña and C., Rehtanz. 2014."Classification of lightning stroke on transmission line using multi-resolution analysis and machine learning." *International Journal of Electrical Power & Energy Systems* 58: 19-31.

[49] Malathi, V., N. S. Marimuthu and S. Baskar. 2010. "Intelligent approaches using support vector machine and extreme learning machine for transmission line protection." *Neurocomputing* 73(10-12): 2160-2167.

[50] Hidayatullah, M. Arif, Iqbal. N, Yusliani and Kanda, Januar miraswan. 2020. *Pengenalan makhraj huruf hijaiyah menggunakan convolutional neural network.* PhD diss., Sriwijaya University. [*Introduction of makhraj hijaiyah letters using convolutional neural network.*]

[51] Pylarinos, D. 2020. "Overhead Transmission Line Easement and Right-of-Way Cases in Crete, Greece: A Statistical Analysis of 1220 Cases from 1974 to 2019." *Engineering, Technology & Applied Science Research* 10(3): 5581-5589.

[52] S. Kannadhasan and R. Nagarajan, Development of an H-Shaped Antenna with FR4 for 1-10GHz Wireless Communications, *Textile Research Journal*, DOI:10.1177/

00405175211003167 journals.sagepub.com/home/trj, March 21, 2021

[53] S. Kannadhasan and R, Nagarajan, Performance Improvement of H-Shaped Antenna with Zener Diode for Textile Applications, *The Journal of the Textile Institute*, Taylor & Francis Group, DOI: 10.1080/00405000.2021.1944523

[54] S. Kannadhasan, G. Karthikeyan and V. Sethupathi, A Graph Theory Based Energy Efficient Clustering Techniques in Wireless Sensor Networks. Information and Communication Technologies Organized by Noorul Islam University (ICT 2013) Nagercoil on 11-12 April 2013, Published for *Conference Proceedings by IEEE Explore Digital Library* 978-1-4673-5758-6/13 @2013 IEEE.

[55] S. Kannadhasan and R. Suresh, EMD Algorithm for Robust Image Watermarking. *Recent Advances in Mechanical Engineering and Interdisciplinary Developments* Organized by Ponjesly College of Engineering (ICRAMID 2014) Nagercoil on 7-8 March 2014, Published for Advanced Materials Research Vols.984-985 (2014) PP 1255-1260, ISSN No:1022-6680

[56] Kannadhasan, M. Isavani and G. Karthikeyan, *A Novel Approach Privacy Security Protocol Based SUPM Method in Near Field Communication Technology, Artifical Intelligence and Evolutionary Algorithms in Engineering Systems (ICAEES 2014)* Organized by Noorul Islam University, Kumaracoil Nagercoil on 23-24 April 2014 Published for Advances in Intelligent Systems and Computing.

In: Research Challenges in Science ... ISBN: 978-1-68507-008-3
Editors: S. Kannadhasan et al.

Chapter 4

ANTIMICROBIAL BEHAVIOUR OF SUBSTITUTED STYRYL 3-NITRO-4-CHLOROPHENYL KETONE

V. Usha*, G. Thirunarayanan[2], V. Thangaraj[3], I. Muthuvel[4] and S. Rajalakshmi[5]

[1]Department of Chemistry,
University College of Engineering Panruti, Panruti, India
[2]Department of Chemistry, Annamalai University,
Chidambaram, India
[3]Department of Chemistry, Anna University, Tiruchirappalli, India
[4]Department of Chemistry, MR Government Arts College,
Mannargudi Department of Physics, India
[5]University College of Engineering Panruti, Panruti, India

ABSTRACT

This article describes the synthesis of substituted styryl 3-nitro-4-chloro phenyl ketones. Molecular formula, molecular weight, melting point, CHN analysis were determined. The existence of *s-cis* and *s-trans* conformers is confirmed by IR spectra. Presence of

* Corresponding Author's E-mail: vushachem2004@gmail.com.

H_α, H_β, ^{13}C of CO, C_α, C_β has been confirmed by NMR spectra. The antibacterial and antifungal activity of all the compounds is determined by disc diffusion method. All the compounds have satisfactory antimicrobial activity. Graphical abstract:

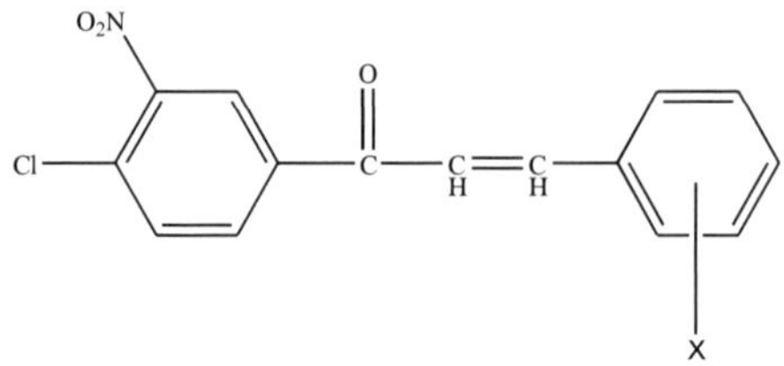

Where X- H, 3-Br, 4-Br, 3-Cl, 4-Cl, 3-F, 3-OH, 2-OCH_3, 4-OCH_3, 3-NO_2

Keywords: Ketone, Crossed Aldol Condensation reaction, IR, NMR, Kirby-Bauer disc diffusion method

INTRODUCTION

Sensible design, synthesis and production are the main objectives of organic chemistry. Many organic compounds are present in nature which are very important in various biochemical fields. Variety of molecules find applications in various fields ranging from drugs, medicines, dyes, etc. Chalcones are the naturally occurring substance that complete the group of flavonoids and to which are attributed to a wide range of biological applications [1]. In medicinal chemistry chalcones and their derivatives play a vital role [2]. The reaction between aldehyde and ketone in presence of base catalyst also known as crossed aldol reaction is very important because of the presence of keto ethylinic moiety. This keto ethylinic moiety is very important for the numerous biological behavior such as antibacterial [3], anti-inflammatory [4], antimalarial [5], antifungal [6], antiviral [7], antioxidant [8], mutagenic [9] and cytotoxic activities [10]. The antibacterial activity of chalcones is being increasingly documented. Sivasankerreddy and co-workers have synthesized diazenyl derivatives having chalcone moiety and *in vitro* screening of the compounds was determined against some bacterial and fungal strain [11]. Design, synthesis, and biological activities of 1-aryl-(3-(2-styryl)phenyl)prop-2-en-1-ones have been performed by Soon Young Shin et al. and concluded that these findings can assist in the design of novel compounds with a combination of the structures of chalcone and resveratrol for use as chemotherapeutic agents [12]. Although a great number of chalcones have been published in the literature, an overview of the recent scientific accounts describing the design of new molecules is still needed. Therefore, this article emphasizes the rationale behind the discovery, the synthesis and antimicrobial activity of the new chalcone derivatives.

RESULTS AND DISCUSSION

IR Spectral Study

Infrared spectra of the styryl 3-nitro-4-chloro phenyl ketone are shown in Figure. 1. The carbonyl stretching frequencies of all the synthesized chalcones in this work are presented in Table 1. The synthesized substituted styryl 3-nitro-4-chlorophenyl ketones exist as *s-cis* and *s-trans* conformers.

Table 1. Infrared stretches ν(cm^{-1}) $CO_{s\text{-}cis}$, $CO_{s\text{-}trans}$ of substituted styryl 3-nitro-4-chlorophenyl ketones

S. No	X	$CO_{s\text{-}cis}$	$CO_{s\text{-}trans}$
1	H	1662	1605
2	3-Br	1662	1608
3	4-Br	1627	1592
4	3-Cl	1659	1600
5	4-Cl	1660	1604
6	3-F	1665	1593
7	3-OH	1655	1590
8	2-OCH_3	1656	1590
9	4-OCH_3	1655	1590
10	3-NO_2	1670	1605

The carbonyl stretching frequency of *s-cis* conformers greater than that of *s-trans* conformer, this is due to the presence of 3-nitro-4-

chlorophenyl group which causes greater strain and they encourage the greater absorption of carbonyl group of *s-cis* conformer than the *s-trans* conformer.

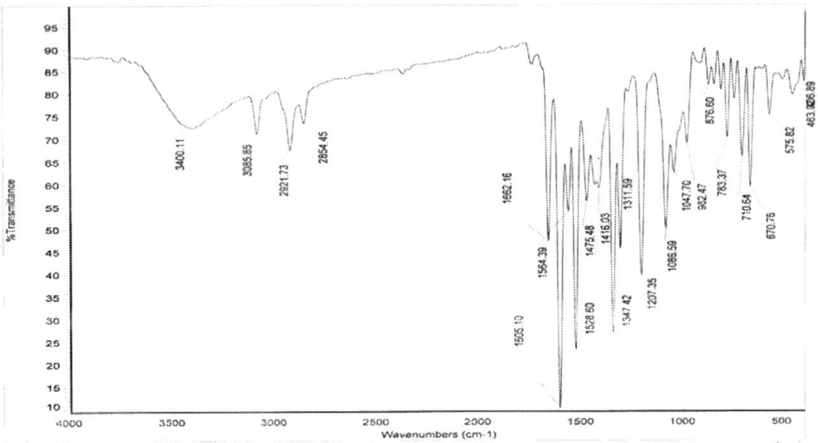

Figure 1. Infrared spectra of styryl 3-nitro-4-chlorophenyl ketone.

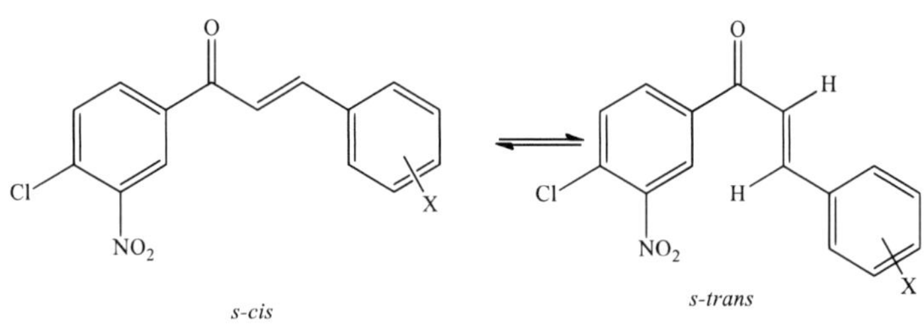

Figure 2. *s-cis* and *s-trans* conformers of substituted styryl 3-nitro-4-chlorophenyl ketones.

These conformers are confirmed by the carbonyl group doublets obtained in the range of 1575-1670cm^{-1}. They are shown in Figure. 2.

^{1}H NMR Spectral Study

The ^{1}H NMR spectra of all the synthesized chalcones have been assigned. NMR spectra of styryl 3-nitro-4-chlorophenyl ketone and its expanded spectra are shown in Figure 3 and Figure 3a respectively.

The assigned proton chemical shift of all the synthesized chalcones are presented in Table 2. The chemical shift of H_α is at the higher field than the H_β because carbonyl group polarize the C=C bond so electron density gets richer at alpha position.

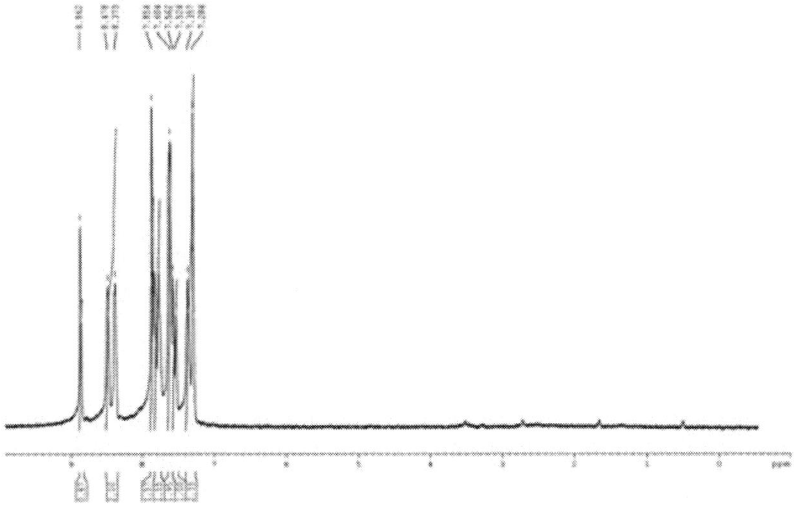

(a)

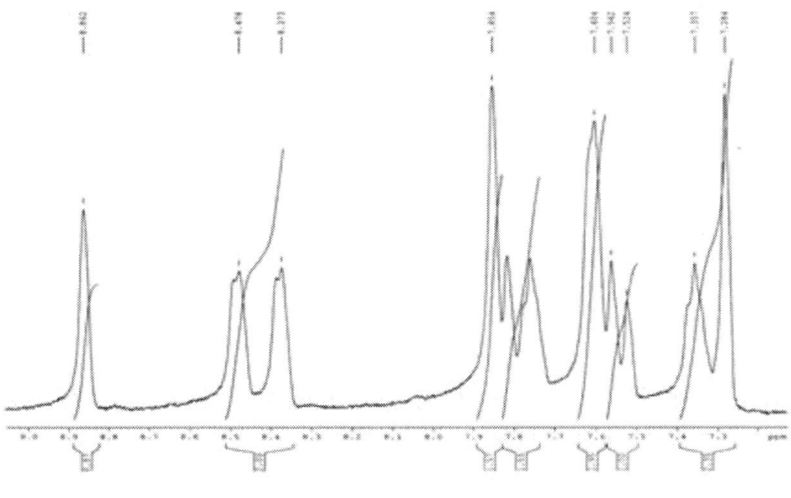

(b)

Figure 3. (a) ^{1}H NMR spectra of styryl 3-nitro-4-chlorophenyl ketone. (b) ^{1}H NMR spectra of styryl 3-nitro-4-phenyl ketone-Expanded.

These chalcones gives doublet in the range of 6-8 ppm. Coupling constant J is calculated as a difference in δ value between doublet and applied frequency. J value is more than 15 Hz.

Table 2. The ^{1}H NMR spectral chemical shifts δ (ppm) of H_{α} and H_{β} of substituted styryl 3-nitro-4-chlorophenyl ketones

Sl. No.	X	Hα	Hβ
1	H	7.548	7.854
2	3-Br	7.560	7.919
3	4-Br	7.049	7.261
4	3-Cl	7.541	7.873
5	4-Cl	6.956	6.999
6	3-F	7.492	7.896
7	3-OH	7.436	7.893
8	2-OCH3	7.630	7.741
9	4-OCH3	7.438	7.896
10	3-NO2	7.688	7.959

^{13}C NMR Spectral Study

^{13}C nmr spectra of the synthesized styryl 3-nitro-4-chlorophenyl ketone are shown in Figure 4. The observed ^{13}C chemical shifts of C_{α} and the C_{β} carbons are presented in Table 3.

Table 3. The ^{13}C Chemical shifts (ppm) of CO, C_{α} and C_{β} of carbons in substituted styryl 3-nitro-4-chlorophenyl ketones

S. No	X	CO	Cα	Cβ
1	H	187.66	121.84	139.22
2	3-Br	188.04	123.29	139.51
3	4-Br	187.67	121.77	139.19
4	3-Cl	187.77	121.04	139.14
5	4-Cl	187.75	121.04	134.13
6	3-F	187.83	120.34	139.46
7	3-OH	187.99	118.28	139.46
8	2-OCH3	187.26	122.06	139.21
9	4-OCH3	188.01	123.22	139.86
10	3-NO2	187.37	122.78	138.90

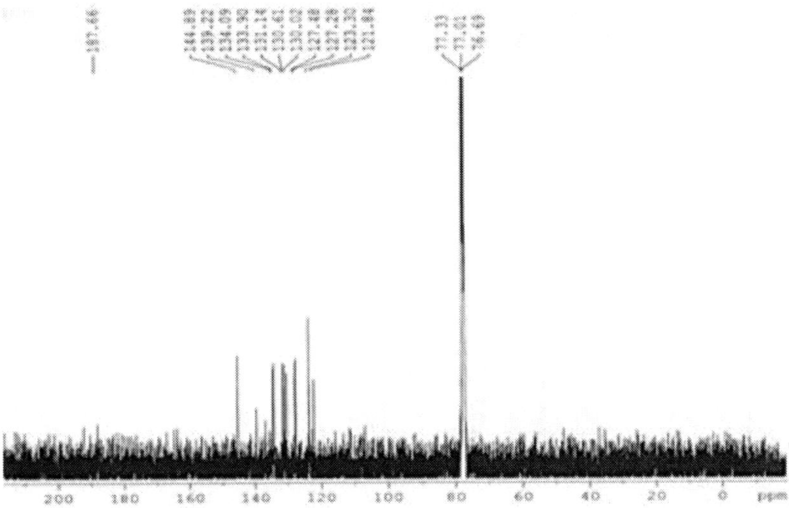

Figure 4. ^{13}C NMR spectra of styryl 3-chloro-4-nitrophenyl ketones.

Antimicrobial Activities

Plants are considered the greatest source to obtain new antimicrobials. Antibiotic refers to the substances produced by microorganisms that act against another microorganism.

Antibacterial Activities of Substituted Styryl 3-Nitro-4-Chlorophenyl Ketones

The evaluation of antibacterial activity was done with *Bacillus subtilis, Micrococcus luteus, Staphylococcus aureus Escherichia coli* and *Pseudomonas aeruginosa*. The antibacterial activity has been done through the Kirby-Bauer disc diffusion technique [13].

The measured antibacterial activities of all substituted styryl 3-nitro-4-chlorophenyl ketones are presented in Table 4. The zone of inhibition of Petri-plates are shown in Figure 5 (Plates 01-10). The clustered column chart has also been done and are shown in Figure 6. The styryl 3-nitro-4-chlorophenylketones 4 and 5 show the best antibacterial activities against *B. Subtilis* strain. Compound 3 show good and 2 show fairly good antibacterial activity and the ketones 6 and 9 show satisfactory and the ketones 1, 7 and 8-10 show the least anti-bacterial activity against *B. Subtilis* strain. The enone compound 5 shows the good and 3, 4 & 6 shows fairly good anti-bacterial activity against *M. luteus*. The ketones 2, 7, 8 and 10 show satisfactory and compounds 1 and 9 show the least antibacterial activity against *M. luteus*. Compounds 4 and 5 show the best antibacterial activity against *S. aureus* strain. Ketone 2 had good antibacterial activity against *S. aureus* strain. The enones 1, 3, 6, 9 and 10 have fairly good antibacterial activity and the compounds 7 and 8 show satisfactory antibacterial activity against *S. aureus*strain. The chalcone 5 had the best antibacterial activity against *E. coli* strain. The chalcones 4show good and the compounds 2, 3, 6 and 10 show fairly good antibacterial activity against *E. coli* strain. Compounds 7-9 showed satisfactorily and 1 show the least antibacterial against *E. coli*. The styryl 3-nitro-4-chlorophenyl ketones 4 and 5 shows the good antibacterial activity and the compound 3& 10 shown fairly good antibacterial activity against *P. aeruginosa* strain. Compounds 1, 2 & 6-8 have satisfactorily antibacterial activity against *P. aeruginosa*. Ketone 9 had the least active against *P. aeruginosa* strain.

Table 4. The mm of a zone of inhibition (mm) values of antibacterial activity of substituted styryl 3-nitro-4-chlorophenyl ketones

		Zone of Inhibition (mm)				
Cpd	X	Gram-positive Bacteria			Gram-negative Bacteria	
		B.subtilis	M.luteus	S.aureus	E.coli	P.aeruginosa
1	H	7	7	9	8	10
2	3-Br	10	9	11	11	11
3	4-Br	11	10	9	10	12
4	3-Cl	12	11	12	12	14
5	4-Cl	12	12	12	13	14
6	3-F	9	10	9	10	11
7	3-OH	6	9	8	9	9
8	2-OCH3	7	9	6	9	10
9	4-OCH3	9	6	9	9	6
10	3-NO2	6	9	9	10	12
Standard	Ampicillin	12	13	12	13	15
Control	DMSO	-	-	-	-	-

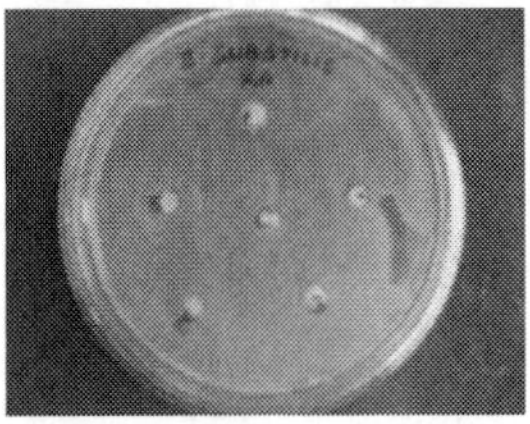

Plate-01

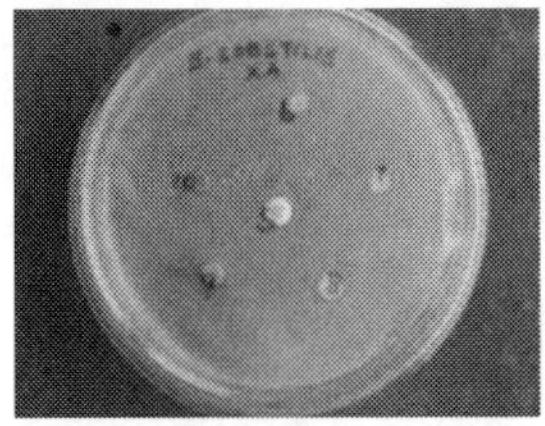

Plate-02

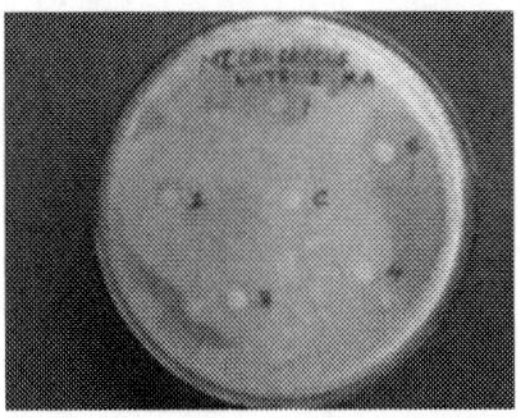

Plate-03

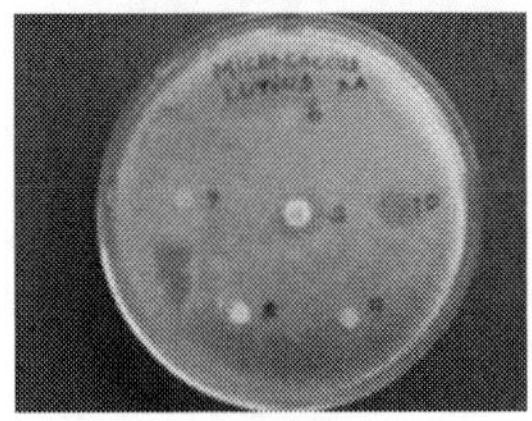

Plate-04

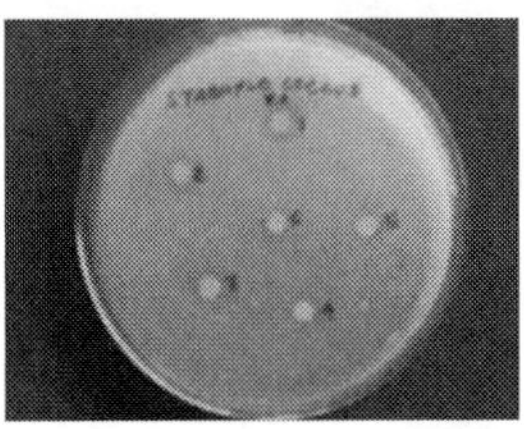

Plate-05

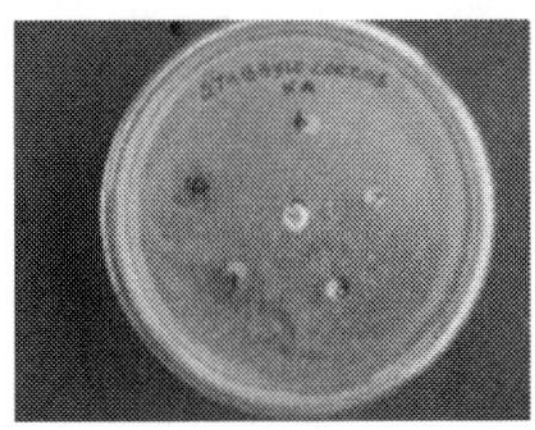

Plate-06

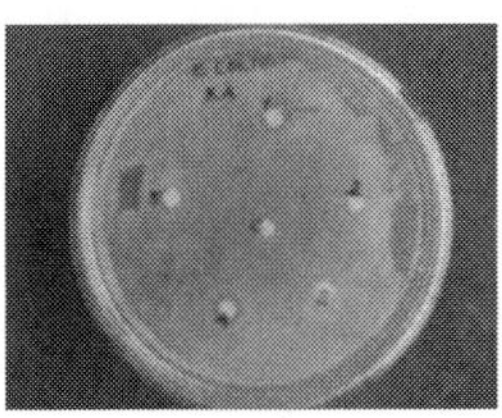

Plate-07

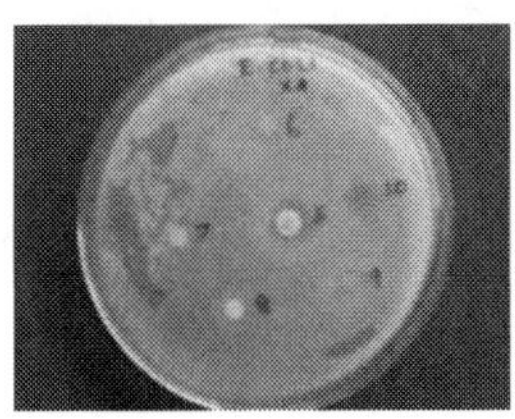

Plate-08

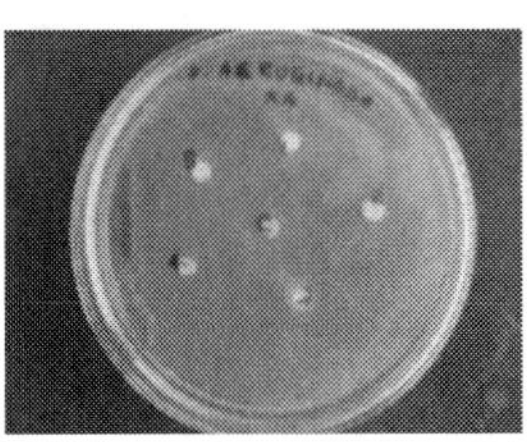

Plate-09

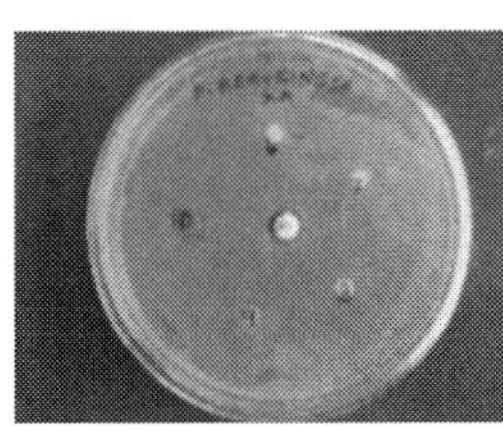

Plate-10

Figure 5. Antibacterial activities- mm of a zone of inhibition of substituted styryl 3-nitro-4-chloro phenyl ketones- Petri plates.

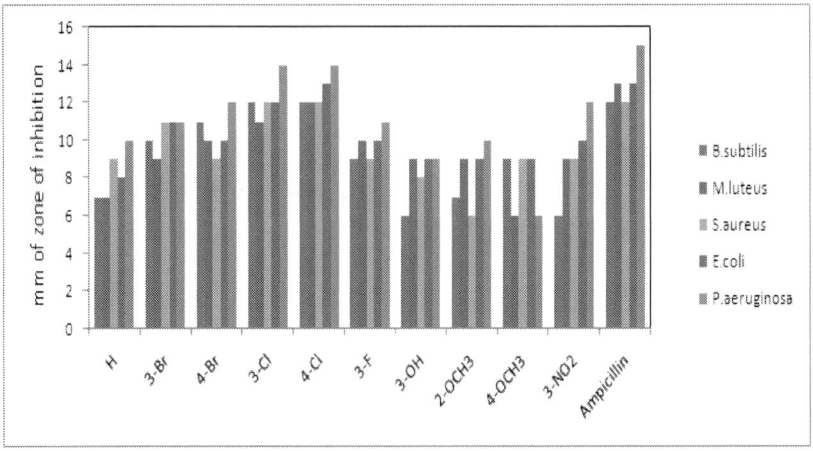

Figure 6. Antibacterial activities of substituted styryl 3-nitro-4-chloro phenyl ketones- Clustered column chart.

Antifungal Activities of Substituted Styryl 3-Nitro-4-Chlorophenyl Ketones

The evaluation of antifungal activities has been done with *Aspergillus niger, Penicilliumscup*, *Mucor species* and *T. viride* fungal strain using the disc diffusion technique [13].

The observed antifungal activities of all substituted styryl 3-nitro-4-chlorophenyl ketones are presented in Table 5. The mm of a zone of inhibition of the compounds against the fungal strains are shown in Figure 7 (Plates 11-16). The clustered column chart of the zone of inhibition is illustrated in Figure 8. The chalcones 9, 10 show good antifungal activity against *A. niger* fungal strain and the chalcones 2, 4, 7 & 8 shows fairly good antifungal activities against *A. niger* fungal strain. The ketones 5 and 6 have satisfactory and enones 1 & 3 have the least antifungal activity against *A. niger* fungal strain. The enone 9 show good antifungal activity against *P. scup*. Compound 10 shows the fairly good antifungal activity against *P. scup* and the ketones 3, 4 and 8 have satisfactory and compounds 1, 2, 5, 6 & 7 show the least antifungal activity. The ketone 10 show good antifungal activity against *T. viride* fungal strain. The chalcones 7 and 9 show fairly good antifungal activities, compounds 1, 2, 3 & 8 exhibits satisfactory antifungal activity against *T. viride* fungal strain. The least antifungal activity was found in the compounds 4-6 against *T. viride.*

Table 5. The zone of inhibition (mm) values of antifungal activity of substituted styryl 3-nitro-4-chlorophenyl ketones

Cpd	X	Zone of inhibition (mm)		
		A.niger	P.scup	T.viride
1	H	5	5	7
2	3-Br	7	5	7
3	4-Br	5	6	6
4	3-Cl	7	6	4
5	4-Cl	6	5	5
6	3-F	6	4	5
7	3-OH	7	5	8
8	2-OCH3	8	6	7
9	4-OCH3	9	9	8
10	3-NO2	9	8	9
Standard	Miconazole	10	10	10
Control	DMSO	-	-	-

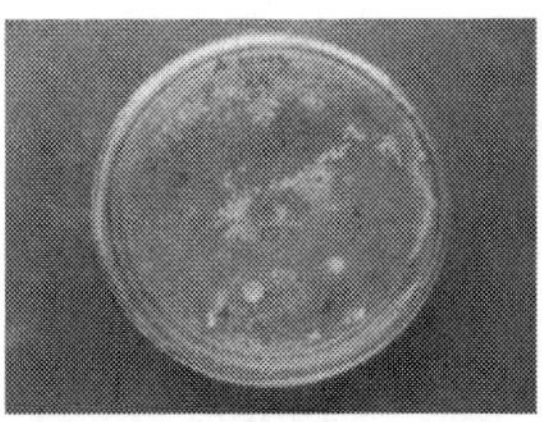

Plate 11

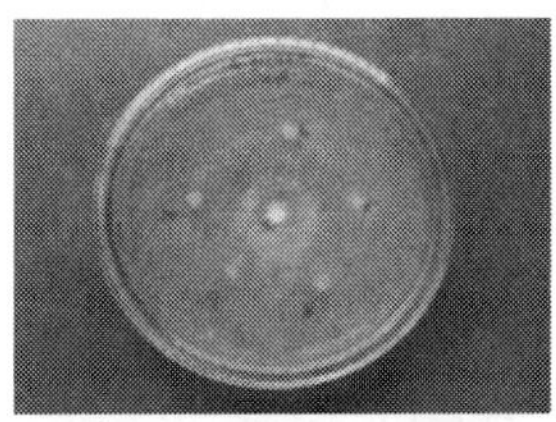

Plate 12

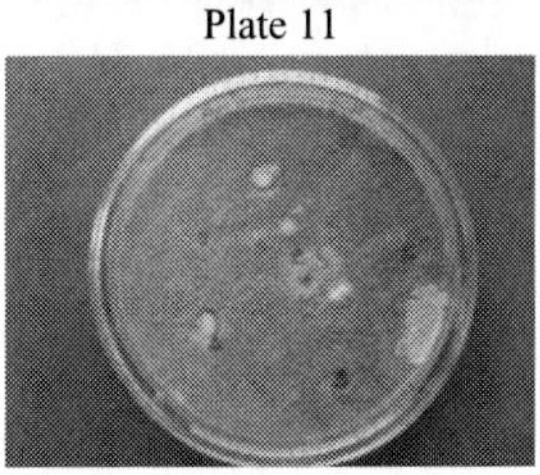

Plate 13

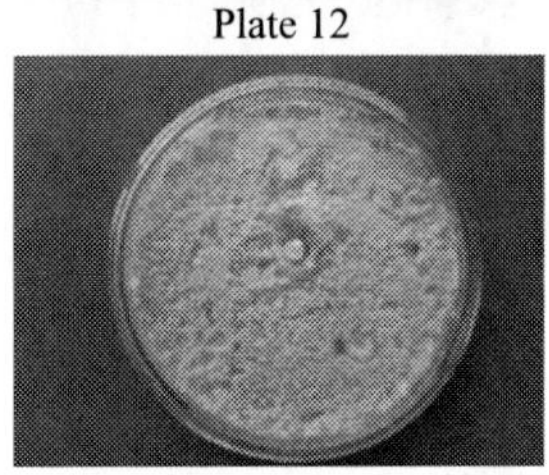

Plate 14

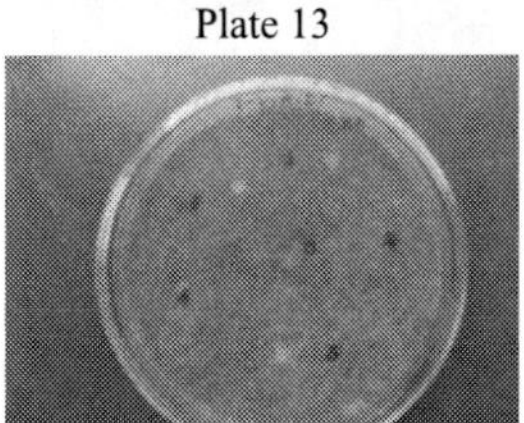

Plate 15

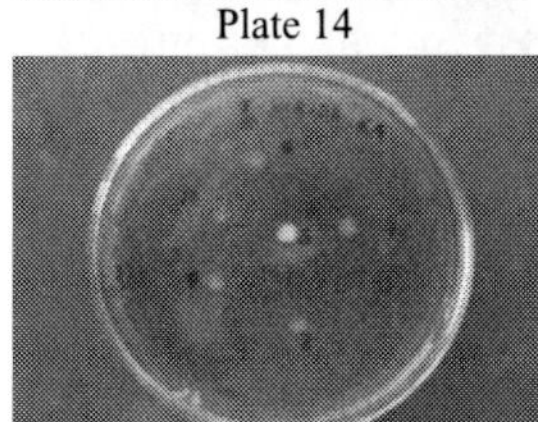

Plate 16

Figure 7. Antifungal activities of substituted styryl 3-nitro-4-chlorophenyl ketones-Petri plates.

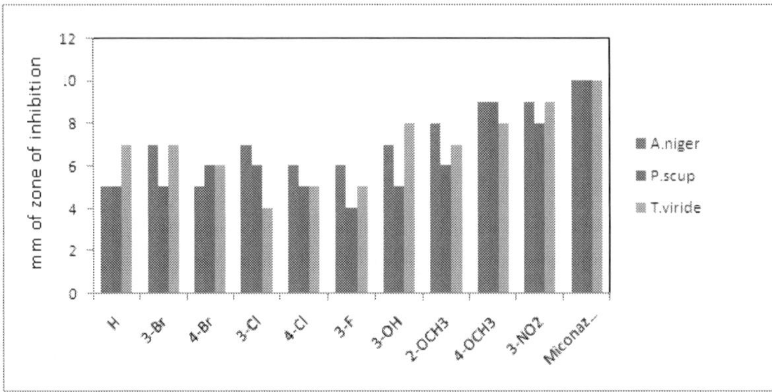

Figure 8. Antifungal activities of substituted styryl 3-nitro-4-chlorophenyl ketones Clustered column chart.

EXPERIMENT

Chemistry

All the reagents and solvents were obtained from commercial companies and used as supplied without further purification. The melting point of all the compounds was determined in an Electrothermal melting point apparatus and is uncorrected. The VARIOMICRO V2.2.0 CHN analyzer was used to carry out the elemental analysis. FT-IR spectra were recorded using Perkin-Elmer FT-IR240C spectrophotometer using KBr0. NMR spectra (^{1}H & ^{13}C) of all the synthesized compounds were performed on BRUKER AVANCE III 400 MHz NMR spectrometer (^{1}H - 400.13 MHz, ^{13}C-100.61 MHz). Antimicrobial activity of all the synthesized compounds was enacted using the Kirby-Bauer disc diffusion method.

Synthesis of Substituted Styryl 3-Nitro-4-Chlorophenyl Ketones

The synthetic method adopted for building a series of ketones were illustrated in scheme 1 - 10. This series of chalcones were prepared by ultrasonicated crossed aldol condensation of substituted benzaldehydes and 3-nitro-4-chloroacetophenone.

Styryl 3-Nitro-4-Chlorophenyl Ketone

An appropriate equimolar quantity of 3-nitro-4-chloroacetophenone (0.398 g, 2 mmol), benzaldehyde (0.212 g, 2 mmol), 10 mL of ethanol and one drop of 1N NaOH were taken in a 25 mL conical flask, closed and ultrsonicated to 15 minutes in room temperature at 40Hz (Scheme 1), (CITIZEN Ultrasonicator, 120W, 40Hz, 240V Ac). The obtained product was filtered by the pump. The product, on recrystallization with ethanol, afforded glittering pale yellow solid, melting at 101-102°C. Yield: 96% (0.55g). Microanalysis for $C_{15}H_{10}ClNO_3$, obtained C, 62.64 (%); H, 3.47 (%); N, 4.82 (%); Req. C, 62.62 (%); H, 3.50 (%); N, 4.87(%); M.W. 287; Mass (m/z): 287 [M^+], 289 [M^{2+}].

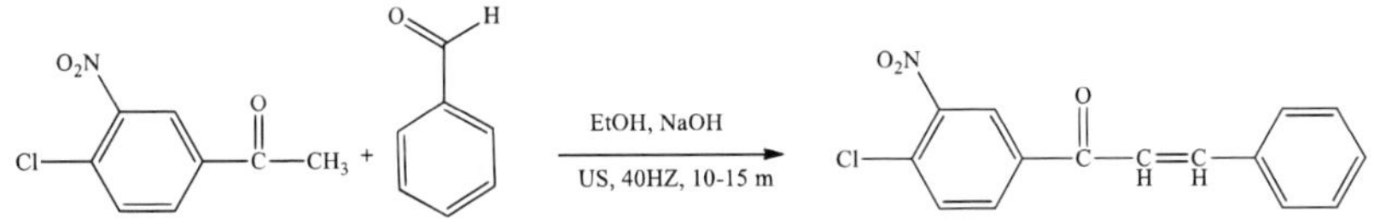

Scheme 1. Synthesis of styryl 3-nitro-4-chlorophenyl ketone.

3-Bromostyryl 3-Nitro-4-Chlorophenyl Ketone

This enone was prepared by following the above procedure with 3-nitro-4-chloroacetophenone (0.398 g, 2 mmol), 3-bromobenzaldehyde (0.378 g, 2 mmol) and 10 mL of ethanol and one drop of 1N NaOH (Scheme 2). Glittering pale yellow solid obtained on recrystallization with etanol. M.P:122-123°C. Yield: 94% (0.69g). Microanalysis for $C_{15}H_9BrClNO_3$, obtained C, 49.17 (%); H, 2.43(%); N, 3.78(%); Req. C, 49.14 (%); H, 2.47 (%); N, 3.82(%); M.W. 366; Mass (m/z): 366 [M^+], 368 [M^{2+}], 370 [M^{4+}].

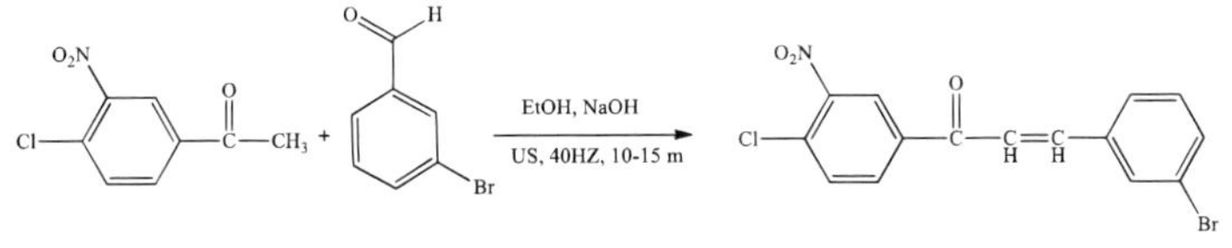

Scheme 2. Synthesis of 3-bromostyryl 3-nitro-4-chlorophenyl ketone.

4-Bromostyryl 3-Nitro-4-Chlorophenyl Ketone

This enone was prepared by following the above procedure with 3-nitro-4-chloroacetophenone (0.398 g, 2 mmol), 4-bromobenzaldehyde (0.378 g, 2 mmol),10 mL of ethanol and one drop of 1N NaOH (Scheme 3). Glittering pale yellow solid obtained on recrystallization with etanol. M.P: 124-125°C. Yield: 94% (0.69g). Microanalysis for $C_{15}H_9BrClNO_3$, obtained C, 49.18 (%); H, 2.41(%); N, 3.78(%); Req. C, 49.14 (%); H, 2.47 (%); N, 3.82(%); M.W. 366; Mass (m/z): 366 [M^+], 368 [M^{2+}], 370 [M^{4+}].

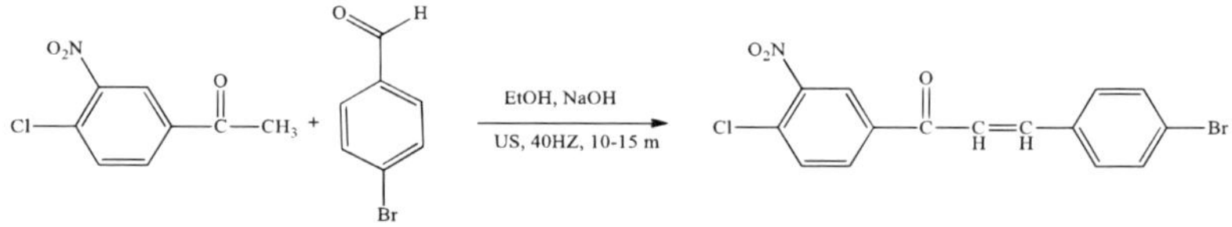

Scheme 3. Synthesis of 4-bromostyryl 3-nitro-4-chlorophenyl ketone.

3-Chlorostyryl 3-Nitro-4-Chlorophenyl Ketone

This chalcone was prepared by following the above procedure with 3-nitro-4-chloroacetophenone (0.398 g, 2 mmol), 3-chlorobenzaldehyde (0.281 g, 2 mmol), 10 mL of ethanol and one drop of 1N NaOH (Scheme 4). Glittering pale yellow solid obtained on recrystallization with etanol. M.P:115-116°C. Yield: 92% (0.59g). Microanalysis for $C_{15}H_9Cl_2NO_3$, obtained C, 55.92 (%); H, 2.76(%); N, 4.32(%); Req. C, 55.93 (%); H, 2.82 (%); N, 4.35(%); M.W. 322; Mass (m/z): 322 [M^+], 324 [M^{2+}], 326 [M^{4+}].

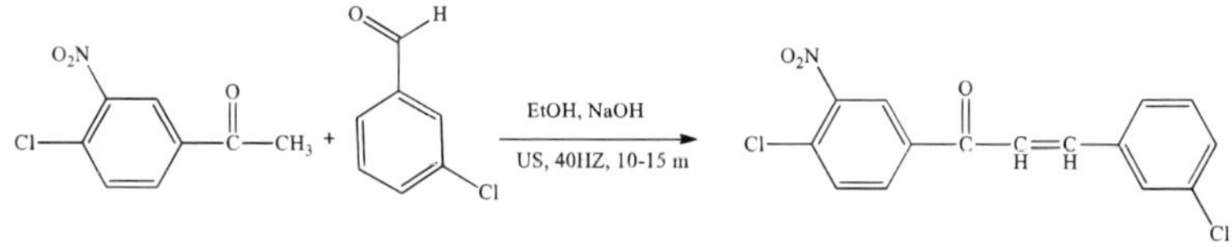

Scheme 4. Synthesis of 3-chlorostyryl 3-nitro-4-chlorophenyl ketone.

4-Chlorostyryl 3-Nitro-4-Chlorophenyl Ketone

This chalcone was prepared by following the above procedure with 3-nitro-4-chloroacetophenone (0.398 g, 2 mmol), 4-chlorobenzaldehyde (0.281 g, 2 mmol), 10 mL of ethanol and one drop of 1N NaOH

(Scheme 5). Glittering pale yellow solid obtained on recrystallization with etanol. M.P: 113-114°C. Yield: 94% (0.61g). Microanalysis for $C_{15}H_9Cl_2NO_3$, obtained C, 55.89 (%); H, 2.78 (%); N, 4.30 (%); Req. C, 55.93 (%); H, 2.82 (%); N, 4.35(%); M.W. 322; Mass (m/z): 322 [M^+], 324 [M^{2+}], 326 [M^{4+}].

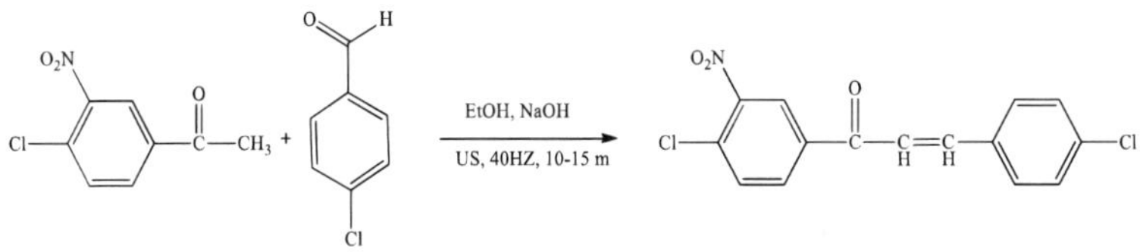

Scheme 5. Synthesis of 4-chlorostyryl 3-nitro-4-chlorophenyl ketone.

3-Fluorostyryl 3-Nitro-4-Chlorophenyl Ketone

This chalcone was prepared by following the above procedure with 3-nitro-4-chloroacetophenone (0.398 g, 2 mmol), 3-fluorobenzaldehyde (0.248 g, 2 mmol), 10 mL of ethanol and one drop of 1N NaOH (Scheme 6). Glittering pale yellow solid obtained on recrystallization with etanol. M.P:127-128°C. Yield: 94% (0.55g). Microanalysis for $C_{15}H_9ClFNO_3$, obtained C, 58.95 (%); H, 2.93 (%); N, 4.54 (%); Req. C, 58.94 (%); H, 2.97 (%); N, 4.58(%); M.W. 305; Mass (m/z): 305 [M^+], 307 [M^{2+}], 309 [M^{4+}].

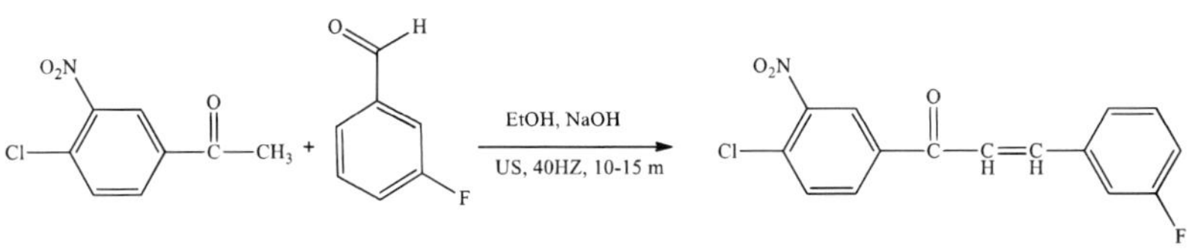

Scheme 6. Synthesis of 3-fluorostyryl 3-nitro-4-chlorophenyl ketone.

3-Hydroxystyryl 3-Nitro-4-Chlorophenyl Ketone

This chalcone was prepared by following the above procedure with 3-nitro-4-chloroacetophenone (0.398 g, 2 mmol), 3-hydroxybenzaldehyde (0.246 g, 2 mmol), 10 mL of ethanol and one drop of 1N NaOH (Scheme 7). Glittering pale yellow solid obtained on recrystallization with etanol. M.P: 122-123°C. Yield: 93% (0.57g). Microanalysis for $C_{15}H_{10}ClNO_4$, obtained C, 59.38 (%); H, 3.28 (%); N,

4.59(%); Req. C, 59.32 (%); H, 3.32 (%); N, 4.61(%); M.W. 303; Mass (m/z): 303 [M^+], 305 [M^{2+}].

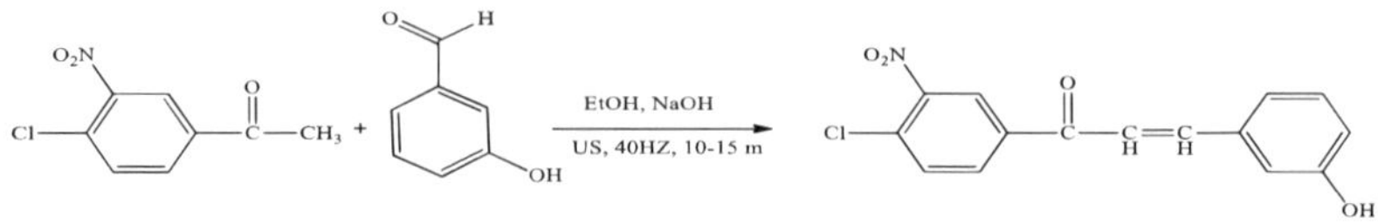

Scheme 7. Synthesis of 3-hydroxystyryl 3-nitro-4-chlorophenyl ketone.

2-Methoxystyryl 3-Nitro-4-Chlorophenyl Ketone

This chalcone was prepared by following the above procedure with 3-nitro-4-chloroacetophenone (0.398 g, 2 mmol), 2-methoxybenzaldehyde (0.272 g, 2 mmol), 10 mL of ethanol and one drop of 1N NaOH (Scheme 8). Glittering pale yellow solid obtained on recrystallization with etanol. M.P:115-116°C. Yield: 95% (0.41g). Microanalysis for $C_{16}H_{12}ClNO_4$, obtained C, 60.53 (%); H, 3.78 (%); N, 4.36 (%); Req. C, 60.48 (%); H, 3.81 (%); N, 4.41(%); M.W. 317; Mass (m/z): 317 [M^+], 319 [M^{2+}].

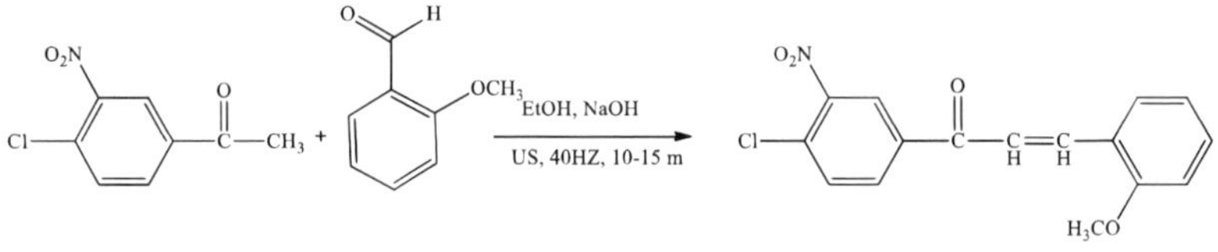

Scheme 8. Synthesis of 2-methoxystyryl 3-nitro-4-chlorophenyl ketone.

4-Methoxystyryl 3-Nitro-4-Chlorophenyl Ketone

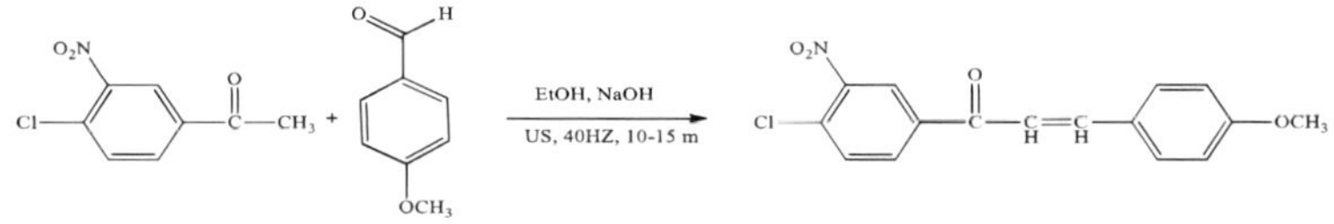

Scheme 9. Synthesis of 4-methoxystyryl 3-nitro-4-chlorophenyl ketone.

This chalcone was prepared by following the above procedure with 3-nitro-4-chloroacetophenone (0.398 g, 2 mmol), 4-methoxybenzaldehyde (0.272 g, 2 mmol), 10 mL of ethanol and one drop of 1N NaOH (Scheme 9). Glittering pale yellow solid obtained on recrystallization with etanol. M.P:125-126°C. Yield: 96% (0.42g). Microanalysis for $C_{16}H_{12}ClNO_4$, obtained C, 60.52 (%); H, 3.79 (%); N, 4.35 (%); Req. C, 60.48 (%); H, 3.81 (%); N, 4.41(%); M.W. 317; Mass (m/z): 317 [M^+], 319 [M^{2+}].

3-Nitrostyryl 3-Nitro-4-Chlorophenyl Ketone

This chalcone was prepared by following the above procedure with 3-nitro-4-chloroacetophenone (0.398 g, 2 mmol), 3-nitrobenzaldehyde (0.302 g, 2 mmol), 10 mL of ethanol and one drop of 1N NaOH (Scheme 10). Glittering pale yellow solid obtained on recrystallization with etanol. M.P:111-112°C. Yield: 90% (0.60g). Microanalysis for $C_{15}H_{12}ClN_2O_5$, obtained C, 54.18 (%); H, 2.68 (%); N, 8.38 (%); Req. C, 54.15 (%); H, 2.73 (%); N, 8.42(%); M.W. 332; Mass (m/z): 332 [M^+], 334 [M^{2+}].

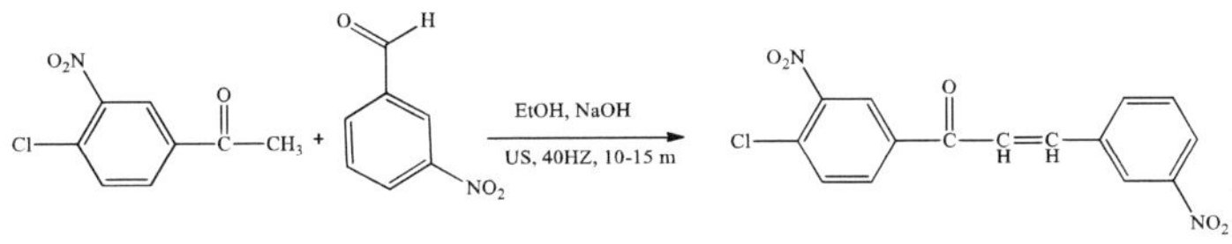

Scheme 10. Synthesis of 3-nitrostyryl 3-nitro-4-chlorophenyl ketone.

CONCLUSION

This method is very efficient for the condensation of substituted benzaldehyde and 3-nitro-4-chloroacetophenone to produce a high yield of 3-nitro-4-chlorophenyl ketone. The synthesized compounds are characterized by their physical constant, microanalysis and spectroscopic data. Halogen substituted compounds produced good antibacterial activities against all the strains. Remaining compounds produced satisfactory antibacterial activity. Methoxy and nitro substituted compounds produced good antifungal activity. The rest of the compounds produced satisfactory antifungal activity.

REFERENCES

[1] Matos, M. J., Vazquez-Rodriguez, S., Uriarte, E., & Santana, L. (2015). Potential pharmacological uses of chalcones: a patent review (from June 2011–2014). *Expert opinion on therapeutic patents*, *25*(3), 351-366.

[2] Amole, K. L., Bello, I. A., & Oyewale, A. O. (2019). Synthesis, characterization and antifungal study of five new derivatives of E-1-(--hydroxyphenyl) chalcone. *Chem Afr* 2: 1–14.

[3] Benouda, H., Bouchal, B., Challioui, A., Oulmidi, A., Harit, T., Malek, F., & Bouammali, B. (2019). Synthesis of a Series of Chalcones and Related Flavones and Evaluation of their Antibacterial and Antifungal Activities. *Letters in Drug Design & Discovery*, *16*(1), 93-100.

[4] Escribano-Ferrer, E., Queralt Regué, J., Garcia-Sala, X., Boix Montañés, A., & Lamuela-Raventos, R. M. (2019). *In vivo* anti-inflammatory and antiallergic activity of pure naringenin, naringenin chalcone, and quercetin in mice. *Journal of natural products*, *82*(2), 177-182.

[5] Hameed, A., Masood, S., Hameed, A., Ahmed, E., Sharif, A., & Abdullah, M. I. (2019). Anti-malarial, cytotoxicity and molecular docking studies of quinolinyl chalcones as potential anti-malarial agent. *Journal of computer-aided molecular design*, *33*(7), 677-688.

[6] Mellado, M., Espinoza, L., Madrid, A., Mella, J., Chávez-Weisser, E., Diaz, K., & Cuellar, M. (2019). Design, synthesis, antifungal activity, and structure–activity relationship studies of chalcones and hybrid dihydrochromane–chalcones. *Molecular diversity*, 1-13.

[7] Patil V, Patil SA, Patil R, Bugarin A, Beaman K, Patil SA. Exploration of (hetero)aryl Derived Thienylchalcones for Antiviral and Anticancer Activities. *Med Chem.* 2019;15(2):150-161.

[8] Stepanić, V., Matijašić, M., Horvat, T., Verbanac, D., Kučerová-Chlupáčová, M., Saso, L., & Žarković, N. (2019). Antioxidant activities of alkyl substituted pyrazine derivatives of chalcones—*in vitro* and in silico study. *Antioxidants*, *8*(4), 90.

[9] Okuno, L., Y., Marumoto, S., & Miyazawa, M. (2019). Antimutagenic activity of flavonoids from Sozuku. *Natural product research*, *33*(6), 862-865.

[10] Ahmadi, S., Mardinia, F., Azimi, N., Qomi, M., & Balali, E. (2019). Prediction of chalcone derivative cytotoxicity activity against MCF-7 human breast cancer cell by Monte Carlo method. *Journal of Molecular Structure*, *1181*, 305-311.

[11] Sivasankerreddy, L., Nagamani, B., Rajkumar, T., Babu, M. S., Subbaiah, N. Y., Harika, M. S., & Nageswarao, R. (2019). Novel Diazenyl Containing Phenyl Styryl Ketone Derivatives As Antimicrobial Agents. *Anti-infective agents*, *17*(1), 28-38.

[12] Shin, S. Y., Lee, J., Park, J., Lee, Y., Ahn, S., Lee, J. H., & Lim, Y. (2019). Design, synthesis, and biological activities of 1-aryl-(3-(2-styryl) phenyl) prop-2-en-1-ones. *Bioorganic chemistry*, *83*, 438-449.

[13] Bauer, A. T. (1966). Antibiotic susceptibility testing by a standardized single disc method. *Am J clin pathol*, *45*, 149-158.

In: Research Challenges in Science … ISBN: 978-1-68507-008-3
Editors: S. Kannadhasan et al.

Chapter 5

COMPUTATION OF CONGESTION COST IN TRANSMISSION LINES

S. Chellam*, PhD, M. Jajini and S. Kalyani, PhD

Department of Electrical and Electronics Engineering,
Velammal College of Engineering and Technology,
Madurai,Tamilnadu,India

ABSTRACT

Transmission Congestion Cost represents the cost associated with re-dispatching of schedules to avoid transmission line limit constraints violations. Re-dispatching of units cause the increase of total generation cost. This paper is an effort to rationalize and systematize the procedures for calculation of transmission congestion cost using tracing based power flow. The power flow in the transmission lines is traced using Power world simulator tool. The increase in real power generation will cause overload at transmission lines. This congested power flow in transmission is employed to determine the congestion cost. The congestion cost is calculated from congested power flow and cost per unit power flow in transmission lines. The power transfer limit has been taken as security constraint while dealing with congestion problem. The proposed methodology is illustrated on IEEE 14 bus test system.

* Corresponding Author's E-mail: scv@vcet.ac.in.

INTRODUCTION

In the deregulated power system, the system operator would like to schedule more generation from the cheapest available sources with the objective of production cost minimization. But the competition among the power producers leads to changes in their bidding methodologies frequently. The scheduling for this producer's strategic bidding leads to allocate high MW quantity at lower price offered utility and low MW quantity at higher price offered utility. By the frequently changing schedules, the possibility unsecured state due to anyone of the transmission line overloading condition will increase. This situation is termed as Congestion in deregulated environment, (Charles Raja. S., Venkatesh, and Manikandan. B.V, 2011) and the congestion cost has been calculated with suitable techniques. With increasing demand for electric power all around the globe, electric utilities have been forced to meet the same by increasing their generation. However, the electric power that can be transmitted between two locations on a transmission network is limited by several transfer limits such as thermal limits, voltage limits and stability limits with the most restrictive applying at a given time. When such a limit is reached, the system is said to be congested [1].

When there is no congestion, increment of loads are allocated to most economical plants. Therefore, zone prices remain same, for each zone as long as the capacities of low cost units remain available. But in the case of congestion, zone prices are increased; become different, and low cost sources remain un-utilized. Therefore, congestion management and proper allocation of its cost has become more significant nowadays in deregulated power systems. This paper (Anup Jana, 2008) quantifies Total Congestion Cost due to congestion, by computing Line wise Impact Cost for each congested line. Using these Line Impact Costs total system congestion cost is apportioned to each congested line [2]. Only most sensitive generators are rescheduled to reduce the computational effort and hence the congestion cost. It has been analyzed that by using PTDF all the other factors can be calculated which are useful in determining the system behavior under different loading conditions. Load and Generator contributions obtained from the PTDF method are utilized to allocate transmission charges using MW Mile method in which magnitude of the power in the transmission line, the length of line and the cost of line are taken into account (A.K. Singh, C. S. Özveren, 2015)

Generator rescheduling and wheeling transactions are considered to optimize the algorithm which is capable of giving optimal solution with less cost in all the cases (A. Ahamed Jeelani Basha, and M. Anitha, 2014) [3]. According to the PFC (Power Flow Comparison) Method, all generators have some contribution to each line flow. Based on their contribution, the order in which the generators should be adjusted can be determined. When a generator (Jian Yang and Max D. Anderson) has a positive contribution to the line flow, the output of this generator needs to be decreased to reduce the transmission congestion. When two generators have conflicts in congestion management, the one with higher MW contribution has more impact on the line flow changes [4]. (Umair Shahzad, 2017) explained the load flow analysis using Newton-Raphson (NR) method with the help of power world simulator which provides less time for computation with accurate results. The re-dispatch of transactions for congestion management in a pool model for single line congestion is formulated as a non linear optimization problem and it has been solved by using DE using MATLAB (F. Hussin, IM.Y. Hassan, K.L. Lo) . In this paper, power flow from generator to transmission line has been traced based on power flow tracing principle. The power tracing is achieved by implementing Power world simulator tool. From the traced power flow, the congestion cost has been calculated. For calculating the congestion cost, the difference in power flow between at base case and congested case is concerned. The proposed method of transmission congestion cost calculation has been implemented on IEEE 14 bus test system. This method of congestion cost computation can be extended for large scale system also.

PROBLEM FORMULATION

The objective function of the proposed method is formulated as,

$$C_{cost} = \Delta P_{gi} * C_{gb} \quad (1)$$

Where,
C_{cost}- Congestion cost
ΔP_{gi} – change in power flow in transmission lines due to generator gi in Mw
C_{gb} – incremental bid cost of generator gi due to power flow in line i and j

Application of Power World Simulator in the Proposed Work

Power World Corporation was formed in 1996 with the goal of further developing and commercializing the Simulator tool. Power World Simulator is an interactive power system simulation package designed to simulate high voltage power system operation on a time frame ranging from several minutes to several days. The software contains a highly effective power flow analysis package capable of efficiently solving systems of up to 250,000 buses. Power flow is a traditional power engineering calculation that is performed to determine the flows on all lines and the voltages at all buses in the system given the power injections at all buses and the voltage magnitudes at some of them. When Simulator performs a timed simulation, it actually performs a sequence of power flow calculations, each based on data that present a snapshot of system conditions at a particular instant. Since it is based on power flow, Simulator models the system as being of constant frequency. This assumption is only an approximation, but frequency deviations are generally very small in everyday operation.

In this paper, power flow between the transmission lines has been traced using power world simulator tool in which convergence occurs quickly. The power flow problem entails solving a system of nonlinear equations. Solving a nonlinear system requires the use of an iterative algorithm to hone in on the correct solution. Many nonlinear system solvers have been developed, and Power World provides access to the full Newton-Raphson (NR) method.

Transmission Congestion Cost Calculation

Transmission congestion occurs when the dispatching of transactions causes the violation on the transmission system. It could be caused for various reasons, such as transmission line outages, generator outages, changes in energy demand and uncoordinated transaction. As a result it prevents the system operators from dispatching additional power from a specific generator. Transmission congestion cost is calculated from the congested flow of power among the transmission lines. For this purpose, the power flow in the transmission line has been calculated for each line. The total generation cost without congestion in all the transmission line is taken as GC. The total congestion cost in all transmission lines is considered as CC. Therefore, the Transmission Congestion Cost in transmission lines is computed as:

$$TCC = CC - GC \tag{2}$$

Where,
TCC - Transmission Congestion Cost
CC - cost due to congestion in lines
GC - generation cost without congestion

Results and Discussion

IEEE 14 bus system, standard test system which includes 5 generators, 9 load positions and 19 branches. In this test system, generator buses are 2, 3, 6 and 8. Bus 1 is considered as slack bus. The test system has the real power generation of 272.4MW and load of 259MW. The power flow from each generator to transmission line has been found by applying tracing principle. The total power flow in all transmission lines without any congestion is 384.29 MW. The congestion in transmission line has been produced by increasing the real power generation at generator buses only. The Power flow in transmission lines can be found using optimal tracing principle which uses power world simulator tool to trace the power flow under normal and congested conditions. Simulated power flow diagram of IEEE 14 bus system is shown in Figure 2.

The real power generation at generator buses is increased. When the power flow limit exceeds in the transmission lines, then overloading (congestion) will occur in the transmission lines. The power flow details due to overloading are given in below:

- Case A : Real power generation is increased at bus 2 as 40% from its base value.
- Congested lines are 1-5 and 4-5 and the total power flow due to congestion is 487.12MW.
- Case B: Real power generation is increased at bus 3 as 65%.
- Congested lines are 2-3 and 4-5; the congested power due to this case is 424.66MW.
- Case C : Real power generation is increased at bus 6 as 70%.
- Congested lines are 1-5 and 2-4; the congested power due to this case is 382.45 MW.
- Case D : Real power generation is increased at bus 8 as 80%.
- Congested line is 7-8 and the congested power due to this case is 436.87MW.

Table1. Congestion Cost Calculation due to % increase in generation limits

Case Study	Congested Lines	Power Flow Due To Congestion (MW)	Cost Due To Congested Power Flow ($/MW)
Case A	1-5,4-5	487.12	9742.4
Case B	2-3,4-5	424.66	8493.2
Case C	1-5,2-4	382.45	7849
Case D	7-8	436.87	8737.4

Table 1 show the cost in the transmission lines due to increase in generation limits at generator buses. Total power flow at each case is shown in table. The cost is calculated by multiplying the congested power flow with the bid rates. The bid rates in all the generator unit is taken as 20$/MW. From Table.2, we can know that 80% increase in generation limit will give a more overloaded power flow in the transmission lines. But the cost in CASE C is 7849$/MW which is a minimum value as compared with other cases.

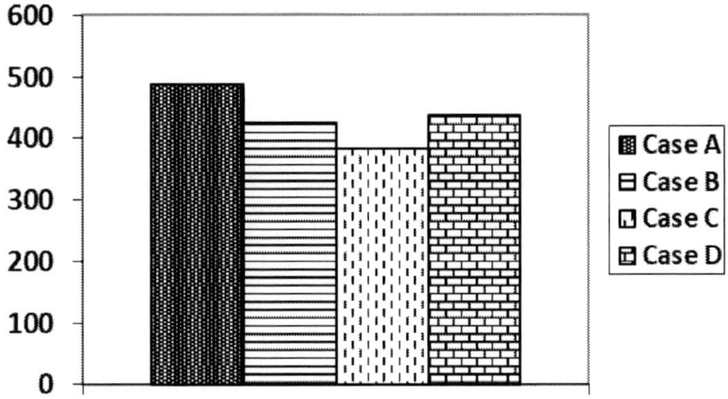

Figure 1 Power flow at base case and case study

The comparison of power flow from base case and congested power flow from the four case studies are shown in Fig.1. The power flow at base case (before congestion) is 384.29MW. The generator bid rate is taken as 20$/MW. Therefore, the cost for the power flow in the transmission is 7685.8 $/MW.

Table 2 Transmission Congestion Cost for various case study

Case Study	Congested PowerFlow (MW)	Cost Due to Congested PowerFlow ($/MW)	Transmission Congestion Cost ($/MW)
Case A	487.12	9742.4	2056.6
Case B	424.66	8493.2	807.4
Case C	382.45	7849	163.2*
Case D	436.87	8737.4	1051.6

The power flow in the transmission lines has been optimized using power world simulator. The cost due to congestion at various conditions has been shown in Table 2.The transmission cost at base case (without congestion) is 7685.8 $/Mw. Here, the transmission congestion cost has been calculated from the difference in power flow. i.e) between base case power flow and power flow at case study. The congestion cost obtained in Case A is more than the congestion cost obtained due to CASE C. Table: 2 show the transmission congestion cost by increasing the real power generation at generator buses. The transmission congestion cost is determined from the base case power flow cost and the cost due to congested power flow. The difference in congestion cost is less in Case C as 163.2 $/MW. In Appendix, the power flow in IEEE 14 bus system using Power world Simulator is shown. Also, the congestion at bus8 with 80% increase in real power generation will produce congestion in transmission line 7-8 which is indicated in red color.

Conclusion

The congestion power flow in transmission line is optimized by increasing real power generation limits at generator buses using Power World Simulator tool. In this paper, the power flow at congested transmission line has been traced using power flow tracing principle. This traced power flow is utilized to calculate the transmission congestion cost. This paper utilized a new approach with simple mathematical formulas to determine the transmission congestion cost. This method of congestion cost calculation has been checked on IEEE 14 bus system. The same procedure can be followed to calculate the congestion cost in high order systems also. This method overcomes the drawbacks of the optimization algorithms by avoiding the premature convergence. The success and effectiveness of this method of calculating the congestion cost is very simple and effective.

References

A.Ahamed Jeelani Basha.A. and Anitha.M.(2004). Transmission Congestion Management in Restructured Power System using

Firefly Algorithm: International Journal of Computer Applications, 85(1)

Abdullah, M. N., NurLyanie Abdullah. and Jamian, J. J. (2017). Optimal Power Generation in Microgrid System using Firefly Algorithm: *International Conference on Electrical Engineering and Informatics*, 25-27 Nov. 2017, Langkawi, Malaysia

Charles Raja. S., Venkatesh. P., and Manikandan. B.V (2011). Transmission Congestion Management in Restructured Power Systems: 2011 *International Conference on Emerging Trends in Electrical and Computer Technology*. (pp. 23-28)

Hussin, F., Hassan, M.Y., Lo., K.L. Transmission Congestion Management Assessment in Deregulated Electricity Market: 4th *Student Conference on Research and Development*, 1-4244-0527-O/06

Jana., A (2008). A New Approach to Transmission Congestion Cost Calculation: *International Conference on Power System Technology and IEEE Power India Conference*, New Delhi: India

Manikuttan, V., Poly Thomas, Emil Ninan, Kariah. S, Prabhakar Karthikeyan and Saintgits. (2017). Optimization of Market Clearing Price Using Firefly and PSO Algorithm in Bilateral Electricity Market: *Proc. Of the 2017 IEEE Region 10 Conference Malaysia*, November 5-8,2017

Rajathy, R. and Harish Kumar. (2012). Power Flow Tracing Based Congestion Management Using Differential Evolution in Deregulated Electricity Market: *International Journal on Electrical Engineering and Informatics*, 4(2)

S. Kannadhasan and R. Suresh. EMD Algorithm for Robust Image Water marking. Recent Advances in Mechanical Engineering and Interdisciplinary Developments Organized by Ponjesly College of Engineering (ICRAMID 2014) Nagercoil on 7-8 March 2014, Published for *Advanced Materials Research* Vols.984-985 (2014) PP 1255-1260, ISSN No:1022-6680

S. Kannadhasan, G.Karthikeyan and V.Sethupathi, A Graph Theory Based Energy Efficient Clustering Techniques in Wireless Sensor Networks. Information and Communication Technologies Organized by Noorul Islam University (ICT 2013) Nagercoil on 11-

12 April 2013, Published for *Conference Proceedings by IEEE Explore Digital Library* 978-1-4673-5758-6/13 @2013 IEEE.

S. Kannadhasan, M.Isavani and G.Karthikeyan, A Novel Approach Privacy Security Protocol Based SUPM Method in Near Field Communication Technology, *ArtificalArtificial Intelligence and Evolutionary Algorithms in Engineering Systems* (ICAEES 2014) Organized by Noorul Islam University, Kumaracoil Nagercoil on 23-24 April 2014 Published for Advances in Intelligent Systems and Computing

S. Kannadhasan and R,Nagarajan, Performance Improvement of H-Shaped Antenna With Zener Diode for Textile Applications, *The Journal of the Textile Institute*, Taylor & Francis Group, DOI: 10.1080/00405000.2021.1944523

Singh, A.K. and Özveren, C. S. (23015). Congestion Pricing In A Deregulated Market Using A.C. Power Transfer Distribution Factors: *International Universities Power Engineering Conference*. 1-4 Sept. 2015, Stoke on Trent, UK

Umair Shahzad. (2017). Power Flow Analysis using Power World Simulator: *D u r r e e s a m i in J o u r n a l*, 3(1)

Verma S, and Mukherjee, (2016). Firefly algorithm for congestion management in deregulated environment: *International Journal of Engineering Science and Technology*, 2016

Yang, J. and Max, D. Anderson. Tracing the Flow of Power in Transmission Networks for Use-of-Transmission-System Charges and Congestion Management: *IEEE transaction*.

In: Research Challenges in Science ... ISBN: 978-1-68507-008-3
Editors: S. Kannadhasan et al.

Chapter 6

HUMAN COMPUTER INTERACTION FOR DETECTION OF HAND GESTURE USING 3D CONVOLUTIONAL NEURAL NETWORK IN MACHINE LEARNING TECHNIQUES

I. Varalakshmi[1,*], D. Mohana Priya[2] and P. Mathivanan[2]

[1]Department of CSE, Manakula Vinayagar Institute of Technology, Puducherry, India
[2]Department of IT, Manakula Vinayagar Institute of Technology, Puducherry, India

ABSTRACT

Human–Computer Interaction (HCI) is a trending technology that creates superlative communication between human and computer. In this chapter, we discuss HCI with hand gesture detection using various machine learning techniques and algorithms. The results are also discussed by comparing various techniques for better understanding. Gesture-based interfaces are used in many applications such as Smartphone's, Televisions

* Corresponding Author's E-mail: varalakshmicse@mvit.edu.in.

(TVs), video gaming, and so on. In this chapter, a novel method for fingertip detection and hand gesture recognition system using an RGB-D camera and 3D convolution neural network (3DCNN). The key frames extracted from static video summarization based on clustering method and 3D CNN algorithm are used as input to improve classification accuracy. The network of processing layers is used to distinguish the hand contour coordinates and detect the fingertip using KCA (K cosine Algorithm). This chapter demonstrates the accuracy and robustness of the interface by evaluating the hand gesture recognition system from camera images. The various existing algorithms such as DC-CNN (Double channel-CNN), K-curvature method, scale invariant feature transform (SIFT) for hand gesture detection have been discussed along with the parameters and results obtained. The calibration of hand position and orientation aims at translating and rotating the hand images to neutral pose and these data are trained by CNN classifier. We provided a validation of the proposed method on recognizing human gestures which shows robust results with various hand positions, orientations and light conditions.

Keywords: HCI, 3DCNN, KCA, K-curvature, gesture detection, SIFT, DC-CNN

1. Introduction

Human–Computer Interaction (HCI) is a trending technology that creates best communication between human and computer. Human–computer interaction (HCI), on the other hand man–machine interaction (MMI) or computer–human interaction (CHI) is the study of interaction between people (users) and computers. It is a discipline concerned with the design, estimate and implementation of interactive computer systems for human use and with the learning of major phenomena surrounding the following line of investigation. Human-Computer Interaction (HCI) has been enormously successful and has deeply changed computing. The basic goal of the HCI is to improve the interaction between user and computers, the respective user needs satisfy the computers. To design systems that diminish the barrier stuck between the human's cognitive model of what they want – to finish and the computer's accepting of the user's task.

1.1. The Importance of the User Interface

- A well-designed boundary and screen is terribly imperative to our users. It is their window to view the capability of coordination.
- It is also the vehicle all the way through which many critical tasks are existing. These everyday jobs often have a direct force on an organization's relations with its customers, and its prosperity.
- A screen's outline and manifestation affect a person in a variety of ways. If they are confusing and inefficient, people will have greater impenetrability in doing their jobs and will make more mistakes.
- Poor design may even chase some people away from a system everlastingly. It can also lead to provocation, disturbance, and amplified stress.

1.2. HCI with Hand Gesture

With the quick advancement of data society, the PC has become an increasingly more significant piece of our day by day life. In the previous few decades, the most well-known means of human-PC communication (HCI) depended on the standard information gadget like the console and mouse. This is fruitful, however the major issue of this procedure is that they are innately absent of effortlessness and truly associating with the PC. This limitation turns out to be clearer with the appearance of new showcase innovation like computer generated experience. Accordingly Gestures are a sort of nature and ability that we as a whole have, so it doesn't need a lot of thought and spotlight on the actual errand as opposed to in transit it collaborates. Motion acknowledgment has become a problem area for as long as a couple of many years. Presently there are two strategies that are utilized principally for motion acknowledgment. One is the utilization of wearable gadgets, for example, hand gloves. The other is based on PC vision. The previous typically gives more dependable information and performs well, yet it expects clients to wear off-kilter gloves that are costly and diminish instinct. The last one relies upon picture handling to extricate highlights of hand from pictures or recordings, which are touchy to natural variables, similar to light changes. As of late, the fast advancement of innovation makes it conceivable to gain more exact information, for example, the development of profundity sensor Kinect. In light of the above investigation, our methodology is in this way to use the Kinect to catch profundity pictures, which are utilized for additional preparation. We have acquainted ourselves with a natural framework with track and perception of the hand act so that clients can apply them to different application scenes, for example, game and homeroom and so on. The commitments made here have wide pertinence. This strategy grants clients to assemble their own motion orders for explicit situations such that they believe is fitting for that activity.

Programmed hand motion acknowledgment utilizing PC vision is significant for different genuine applications like Sign Language Recognition (SLR). As of late, because of the far and wide accessibility of savvy computerized cameras, investigation into communication via gestures acknowledgment has expanded drastically. Notwithstanding, hand motion acknowledgment actually experiences difficulties because of elements, for example, intricacy of hand signal constructions, and contrasts close by size and hand pose, just as variety of in light and foundation, which can impact the exhibition of hand motion acknowledgment calculations. Variety in motions additionally exists, since individuals can sign a similar hand motion in an unexpected way; undoubtedly a similar individual may sign a similar hand motion in an unexpected way. Moreover, vision-based hand motion acknowledgment is likewise vulnerable to ecological factors like recognizability of the hand. Past ways to deal with hand signal acknowledgment utilizing old style vision calculations ordinarily comprise two fundamental advances. The motivation behind the initial step is to remove highlights, and the reason for the subsequent advance is to apply an AI classifier for playing out the characterization task. Vigorous and successful element portrayal is a significant issue. The previous hand tailored component as a rule requires the client to have some earlier information and to carry out some preprocessing methods, for example, picture change and division to upgrade information quality, separate the pertinent parts and set up the information for the acknowledgment interaction. The yield of the preprocessing stage is spotless information that can be utilized straight forwardly and effectively in the component extraction measure; the nature of pre-handling influences the accomplishment of any acknowledgment technique. Nonetheless, customary hand-planned element extraction procedures are drawn-out and tedious, and can't deal with crude pictures, in contrast with programmed highlight extraction strategies by which helpful highlights can be recovered straightforwardly from pictures

Motions and Gesture Recognition terms are progressively experienced in human PC communication, communication via gestures and PC vision. Motions are the actual activities that pass on some significant data. Motion acknowledgment frameworks as another info methodology in human PC communication give a natural and more advantageous method of cooperation. On the off chance that we disregard the universe of PCs and think about human-human cooperation for a period, we can undoubtedly understand that we are using an expansive scope of signal in close to home correspondence.

The huge utilization of motions in day by day life spurs the utilization of gestural interface in present day time. Hand signal has gotten an incredible decision for communicating the basic thoughts, deciphered by motion acknowledgment framework and transforms them into related occasions. Diverse hand motions that are ordered on the bases of exceptional hand shapes, example, direction or finger positions, can possibly be associated with the PC framework to pass on significant data for human PC cooperation. It is the interest of cutting edge innovation to perceive, characterize and decipher different basic hand signals and use them in wide scope of utilization through PC vision.

1.3. Types of HCI

Theoretically, there are two types of hand gestures in the literature: static gestures and dynamic gestures. The definition of static hand gestures is the appearance characteristics of the hand in the space for a period of time without any movement and dynamic gestures include hand movement information over a period of time, such as waving of the hand, while static hand gestures do not include movement like using the forefinger and ring finger to form the "victory" pose is a static posture. In the scope of this research, we only consider static gestures.

2. Related Work

In early there were many gesture recognition techniques have been developed for recognizing and tracking various hand gestures i.e., Instrumented gloves, optical markers and some advanced techniques based on image features, color based, vision based techniques are available for hand gesture recognition, all have their advantages and drawbacks. An instrumented glove contains a number of sensors which gives information about hand location, orientation and tips of fingers.

These data gloves provide high accuracy results but they are too expensive and require wired connection with the system. Optical tracking systems based on markers work with infrared (IR) light. The markers project IR light and reflect this light on the screen. These systems require a complex configuration.

Recently, some novel techniques have been developed to solve the HCI bottleneck. This difficulty has been largely considered and many feasible solutions have been found. Among these approaches, gestural interaction stands out from other methods, as people usually depend on their hands to interact with the environment. Therefore, it provides a natural and immersive experience to users and gesture-based interaction has many advantages over traditional interactive mode.

Image Based techniques require processing of image textures or color features. If we are working on skin color based detection, there is a wide variation in skin tones from one continent of earth to another. So here comes a simple shape based approach for hand gesture recognition as almost every person possesses the same hand shape with four fingers and a single thumb. The method described in this chapter for gesture recognition requires compactness calculation, which is one of the shape based descriptors and it is calculated as the ratio of squared perimeter of the shape to its area. That means if two different hand shapes with the same perimeter to area ratio exist, it classifies these two different shapes as the same. This method can distinguish only 10 patterns of hand shapes but the proposed method in this paper can recognize any number of the hand shapes based on their orientation and generate different encoding bits with minimal computation time. The algorithm is divided into four main steps, the first step includes pre-processing and segmentation of hand image to reduce the noise from captured input image. In the second step it detects the orientation of hand shapes to categorize them into two categories, one is horizontal and other is vertical. In the Third step it calculates the crucial features required for generating the bit series for different hand shapes and finally recognition and understanding of different hand gestures.

3. Hand Gesture Recognition Model

The structure of the gesture recognition model introduced in this study appears in Figure 1. As shown in the figure, the model consists of three stages: preprocessing, feature extraction, and pattern classification. In the preprocessing phase, to obtain only the object region of the input image, background segmentation is performed by a fast and simple skin color detection technique.

The first part of the feature extraction stage extracts the motion information from the sequence of images by comparing each frame with its preceding frame. After the spatiotemporal template is constructed, the volume data is fed to the feature extractor. Three-dimensional feature data within the receptive fields are then processed by the three-dimensional processing unit in the modified CNN model. A set of feature maps is generated based on the motion factor.

The feature maps generated from the feature extractor are used as the input of the pattern classifier. Since the dimension of the feature face is very high for three dimensional data, it is computational demanding. Therefore we are adopting a feature selection technique to improve the efficiency of the recognition system. The process of the FMM model based feature selection reduces the number of feature types (dimension). The techniques enable the real time recognition without regenerating the performance of the classifier.

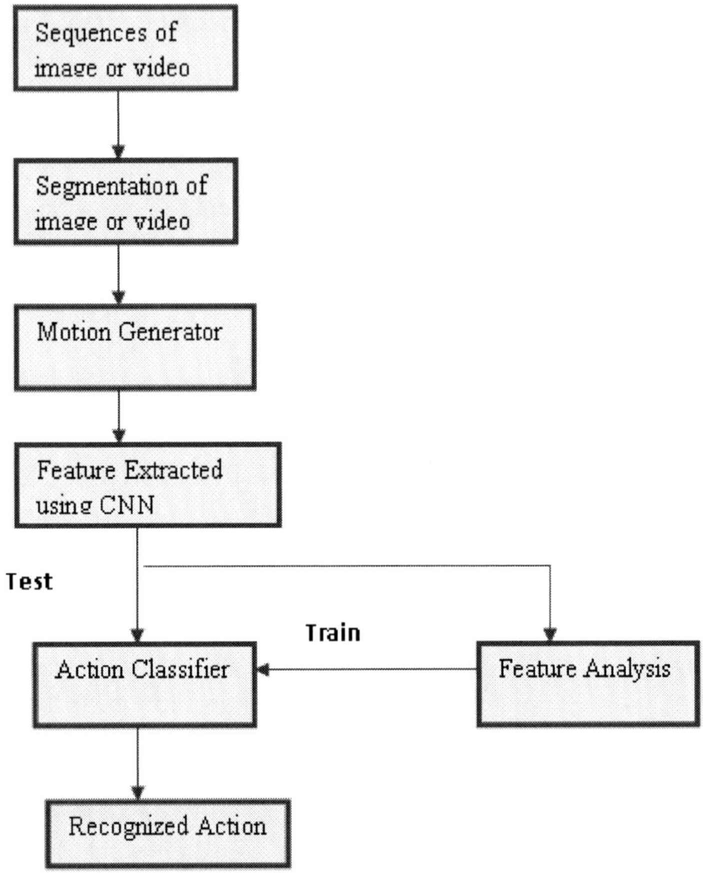

Figure 1. Overview of proposed hand gesture recognition model.

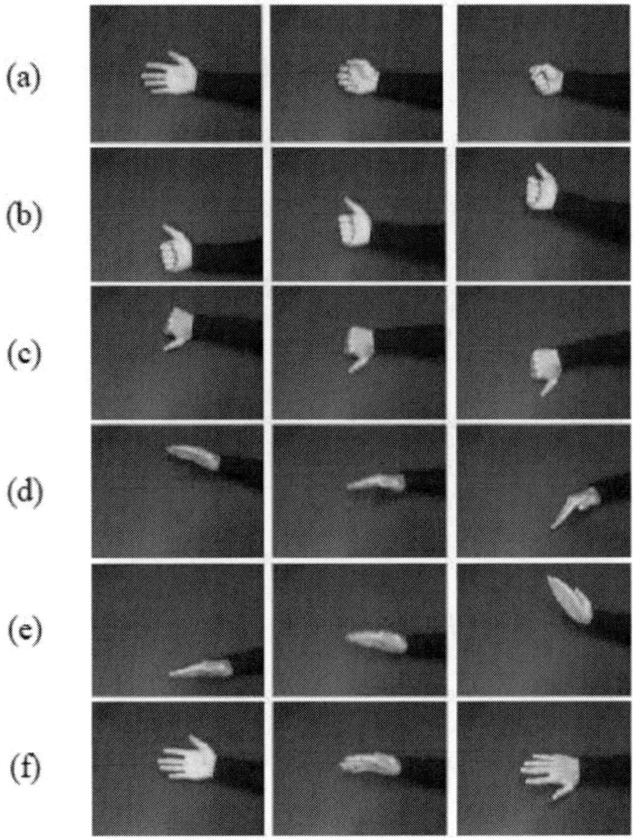

Figure 2. Example of Dynamic Hand Gesture.

In this above mentioned figure have explained the six types of the dynamic hand gesture recognition types. The first hand gesture is Grasp, second one is Thumb up, third one is Thump down, Fourth one is Wave Down, fifth one is Wave up and last one is Turn over.

This section provides background to the CNN, a description of the Convolutional Neural Networks (CNN), baseline CNN (CNN), and a description of the Adapted CNN (ADCNN) architectures, double-channel (CNN).

3.1. Convolutional Neural Network

Convolutional Neural Networks have attained success in image classification problems. The CNN contains three types of layers; namely, convolutional layer, sub sampling or pooling layer, and fully connected layer. Normally, the entire CNN architecture is obtained by stacking several of the above-mentioned layers. The features extracted from a CNN using the three types of layers are hierarchical. The bottom layers in a CNN collect the low level features and high-level layers collect and learn features with more abstract information; this is useful for classification tasks. A convolution layer investigates the spatially-local correlation of its input and maps it into the next layer, which is called a feature map. Different feature maps are constructed from different kernels. The activity of the ith feature map in the lth layer is calculated by (1).

$$y_i^l = \sum_j f(w_{i,j}^l * y_j^{l-1} + b_i^l)$$

where y_i^l is the ith feature map in the lth layer. $W^l_{i,j}$is the convolutional kernel (weight matrix) for$\mathbf{y}^{l-1}_j$, and the weight matrix connects $\mathbf{y}^{l-1}_j$to the feature map $\mathbf{y}^l_i$, $\mathbf{b}^l_i$ is bias of the ith feature map in the lth layer, and f (.) is a nonlinear activation function, such as the rectified linear units (ReLU) function or sigmoid function. The '*' is denoted as the convolutional operator. The first convolutional layer can be followed by the pooling layer. The function used in the pooling layer is the max function or average function, and the most common function used in this layer is the max pooling function. This max function computes the high-level feature in a local window. The polling layer is used to reduce the size of the features and to decrease the required computation time.

3.2. Baseline CNN Architecture: CNN

The baseline deep CNN design is composed of two convolutional layers, two pooling layers and two fully connected layers with ReLU (Rectified Linear Unit). Three dropout performances are in the network (see Table I, Figure 2). There are two dropout procedures after the two pooling layers, and the third dropout is performed after the first fully connected layer. The baseline structure of the hand gesture recognition model developed in this study is described in Figure 1. As shown Figure 1, the model consists of two stages: 1) preprocessing in the first stage; and 2) pattern classification using CNN deep learning algorithm with two different architectures in the second stage.

The first convolutional layer has a part size of 5x5 pixels that contains 32 element maps with a ReLU enactment work. This layer takes pictures with 128×96 pixel esteems as information (see Figure 2). The following layer is a Max-Pooling layer that is designed with a pool size of 2×2. The Max-Pooling layer utilizes the greatest worth to continuously decrease the spatial size of the portrayal to lessen the measure of boundaries and calculation in the organization, and henceforth to likewise control overfitting. The Max-Pooling layer extricates sub locales of 2×2 of the component map, keeps their greatest worth and disposes of any remaining qualities. The following layer is a regularization layer (Dropout) that was designed to haphazardly bar 20% of neurons to lessen overfitting. The following secret layer is a convolutional layer that has a portion size of 3×3 pixels and furthermore contains 32 element maps with a ReLU enactment work.

This layer is trailed by another pooling and regularization layer that is equivalent to the past layer. A while later, a layer called Flatten changes the two-dimensional grid information over to a vector, consequently permitting the last yield to be prepared by standard completely associated layers to acquire the following layers. The principal completely associated layer with the ReLU initiation work contains 128 neurons. This is trailed by a dropout layer to avoid 20% of neurons to decrease over fitting. The second completely associated layer containing 64 neurons with the ReLU enactment works by getting the past 128-dimensional yield of the principal completely associated layer. The last piece of the CNN structure is the yield layer which includes a Softmax initiation work, and contains 6 neurons, one for each hand motion acknowledgment class. The yield is the planning of the information to the last classes for hand signal acknowledgment. Figure 3 addresses the CNN technique.

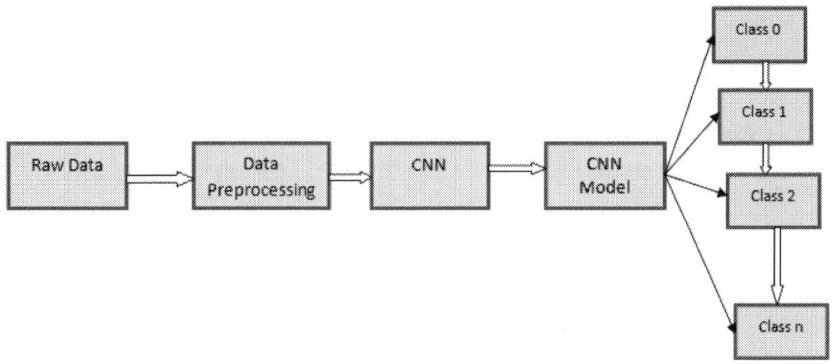

Figure 3. Overview of the proposed hand gesture recognition model.

Table 1. Baseline CNN configuration

Layers Operation	Layers Configuration
Convolution	32 filters, 5x5 kernel and ReLU
Max-Pooling	2x2 kernel
Dropout	20%
Flatten layer	800 Neurons
Fully connected	128 Neurons
Dropout	20%
Fully connected	64 Neurons
Output layer	Softmax 6 classes

3.3. Adapted CNN Architecture: ADCNN

In this section the fine-tuning and adaptation techniques applied to the basic CNN architecture are discussed. This fine-tuned and adapted model, ADCNN, is shown in Figure 4. The performance of the baseline CNN was improved by tuning the parameters to include network initialization and regularization. The proposed approach also includes data augmentation.

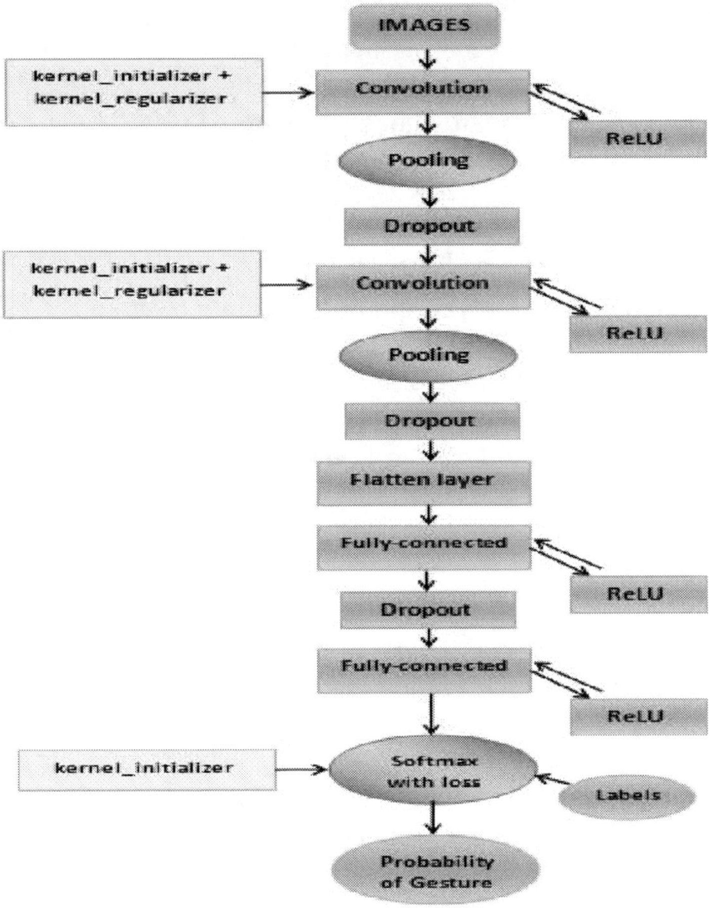

Figure 4. Architecture of the adapted deep CNN (parameters fine-tuning) for hand gesture recognition (ADCNN).

3.3.1. Data Augmentation

It is realized that the more information an AI calculation approaches, the more compelling it may be. In any event, when the information is of lower quality, calculations can really perform better, as long as valuable information can be separated by the model from the first dataset. Information expansion changes the base information, and builds the quantity of preparing information. The changed pictures are normally delivered from the first pictures with next to no calculation and are created during preparing. This implies that preceding taking care of information to the organization for preparing, a change in interaction on the preparation information is applied. Such changes permit the organization to straightforwardly notice the impacts of applying them on information and to prepare the organization to act better on such models. The objective isn't just to decrease over-fitting by means of expansion, yet additionally to increase information to improve the classifier. With respect to ADCNN, arbitrary flat and vertical movements were applied to the information to infer new pictures. Specifically, utilizing information expansion the pictures are moved both evenly and vertically to a degree of 20% of the first measurements arbitrarily, to mathematically build the size of the dataset and to add the heartiness required for a profound learning approach. The changed pictures are added to the dataset to help the CNN execution, forestall over-fitting, and furthermore empowered the calculation to gain better from an assortment of pictures about a similar signal.

3.4. Double-Channel (CNN)

The multi-channel of CNN is utilized to get various perspectives on the information, (for example, red, green, blue channel and sound system sound track of shading pictures). It empowers CNN to learn more copious highlights and the order impact of the model is improved by adding contribution to grouping.

Thus, to advance the preparation interaction of motion acknowledgment, the twofold channel CNN (DC-CNN) is proposed in this section, and the design appears in Figure. The DC-CNN structure is made out of two moderately self-deciding convolution neural organizations. The contribution of the first CNN is the sign picture subsequent to preprocessing, and the info second CNN is motion edge picture. Each one channel contains a similar number of convolutional layers and boundaries, yet they make them decide loads. After the pooling layer, the twofold channels are separately associated with a full association layer and a full association map is performed. The two channels are associated with a completely associated covered up layer, which delivers the yield of a consistent relapse classifier. The substantialness of each channel has its own meaning. However, the last mistake is made entirely through two yield layers. Thus, two yield layers resemble a deviation from one another. This delivers a particular channel tuning dependent on the edge improvement of the Canny channel. The preparation strategy for this model is forward computation, at that point back spread blunder update weight. During the time spent forward computation, to speed up the preparation speed and decrease over fitting, the irregular angle drop (SGD) technique was embraced to ascertain.

4. 3D Convolutional Neural Networks

This system consists of two main phases: 1) video preprocessing and 2) feature learning and classification.

4.1. Video Preprocessing

In this phase, the input videos were converted into sequences of RGB frames of different lengths. To apply 3D CNN on the input sequences, we need to unify the spatial and temporal dimensions of all sequences. For that reason, we started by resizing each frame to a fixed size of 112 x 112 pixels to reduce the computation cost of the successive steps. Next, a linear mapping was used to unify the length of the frame sequences. Whatever the length of an input sequence, it is linearly mapped to a 16 frames sequence. The indices of the selected frames from the input sequence are calculated as

$$index = round\left(\frac{len(input)}{16}\right) * i$$

where len(input) is the length of the input sequence and i = 1 : 16.

4.2. Feature Learning and Classification

In this phase, we employed transfer learning to overcome the lack of a huge annotated gesture dataset. We initialize the 3D CNN with the C3D structure trained on the large-scale Sport-1M2 human action recognition dataset. The investigated model in this work contains six blocks as illustrated in Figure 6.

The primary square comprises a 3D-CNN layer (conv1) of 64 portions followed by a maximum pooling layer. The subsequent square contains a 3D-CNN layer (conv2) of 128 pieces followed by a maximum pooling layer. The third and fourth squares each incorporate two progressive 3D-CNN layers (conv An and conv B) and a max pooling layer. The fifth square incorporates two progressive 3D-CNN layers (conv An and conv B), a zero-cushioning layer and a maximum pooling layer. The quantity of bits in the 3D-CNN layers of these three squares is 256, 512, and 512, separately. The 3D-CNN parts have a brought together state of (3 x 3 x 3) through the entire framework. The portion size of all pooling layers are (3 x 3 x 3) with the exception of the one in the primary square, which has a size of (1 x 2 x 2). The last square contains two completely associated layers every one of 4096 neurons and the yield layer. In completely covered up layers of the model, a redressed straight unit is utilized for enactment, which is the least difficult nonlinear capacity to be utilized for actuation. We attach a softmax prescience layer to the last completely associated layer. The softmax enactment work yields the likelihood for every one of the classes. The anticipated class is the single with the most noteworthy chance. The depicted model was adjusted by back-engendering with negative log-probability to address the hand motions acknowledgment issue.

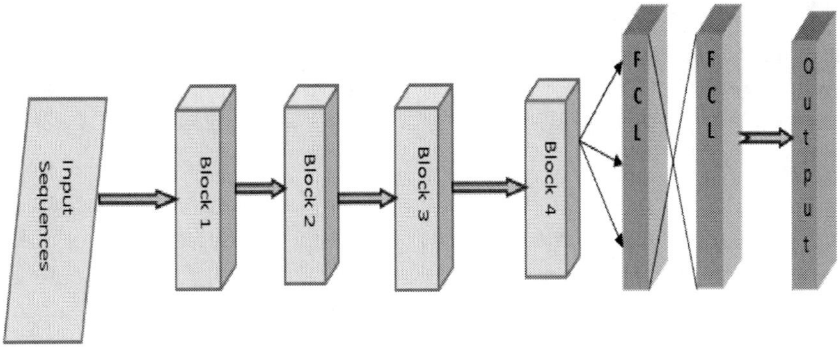

Figure 6. Feature learning and classification for 3D-CNN model.

The notable stochastic inclination plunge analyzer (learning rate = 10-4, rot = 10-6, and energy = 0.9) is utilized when fitting the model available signal datasets. The stacked squares of 3D CNN address the spatiotemporal highlights of the info cut as 3D vectors in various degrees of reflection. The yield of the last 3D-CNN layer was smoothed and sent to the completely associated layers and afterward to the yield layer. To accomplish great speculation for the model on concealed information and stay away from overfitting, we utilized dropout with a likelihood of 0.5 after each completely associated layer.

Conclusion

Concerning the future work, we will sum up the assessment of the proposed framework on more hand signal classes. We will likewise research the impact of various boundaries on the framework execution like the casings' determination strategy, the improvement technique, and the misfortune work. The proposed framework utilized straight planning for outline choice. The exhibition of the proposed framework has been contrasted and some other best in class strategies in the writing. There is still a ton of room for innovative work of DC-CNN, including the accompanying three angles. (1) Try to present more highlights of pecking order and scale to additionally improve the model versatility to complex foundation. (2) The pace of dynamic signal acknowledgment actually has a lot of room for development, and the model can be applied to the field of dynamic motion acknowledgment. (3) The convolution neural organization model for signal acknowledgment requires a ton of picture information with marks for preparing. Later on, preparation can be brought out through unaided learning or semi-directed figuring out how to decrease the reliance of the model on countless label information. In choosing a successful element to diminish dimensionality, the importance factor has been utilized. Along these lines, the procedure improves the calculation speed and heartiness to commotion. Further exploration incorporates an improved acknowledgment model utilizing composite data, investigates more intricate hand motions. The proposed ADCNN was contrasted with a benchmark CNN. The outcomes uncovered that the proposed approach which incorporates information increase, alongside a CNN tuned network utilizing network introduction (ReLU and Softmax), and L2 regularization accomplished the most elevated characterization acknowledgment precision 99.73% - a 4% improvement over the benchmark CNN model which was not tuned utilizing the previously mentioned boundaries.

References

Aashni Haria, Archanasri Subramanian, Nivedhitha Asokkumar, Shristi Poddar. & Jyothi S Nayak. (2017). "Hand Gesture Recognition for Human Computer Interaction" *7th International Conference on Advances in Computing & Communications*, Cochin, India.

Ali A. Alani, Georgina Cosma. & Aboozar Taherkhani. (2018). "Hand Gesture Recognition Using an Adapted Convolutional Neural Network with Data Augmentation" *4th IEEE International Conference on Information Management*, 978-1-5386-6147-5/18/$31.00.

Alper Yilmaz. & Mubarak Shah. (2015). Actions Sketch: A Novel Action Representation. *IEEE Computer Society Conference on Computer Vision and Pattern Recognition*, Vol. *1*, (2005), 984-989.

Assaleh, K., et al., (2010). "Continuous Arabic sign language recognition in user dependent mode," *J. Intell. Learn. Syst. Appl.*

Bao, P., Maqueda, A. I., del-Blanco, C. R. & Garcıa, N. (2017). *"Tiny hand gesture recognition without localization via a deep convolutional network," IEEE Trans. Consum. Electron.*, vol. *63*, no. 3, pp. 251–257.

Cristophe Garcia. & Manolis Delakis. (2004). Convolutional Face Finder: A Neural Architecture for Fast and Robust Face Detection. *IEEE Transactions on Pattern Analysis and Machine Intelligence*,. Vol. *26*, No. 11, (2004), 1408-1423.

Gongfa Li, Heng Tang1, Ying Sun, Du Jiang, Bo Tao, Shuang Xu. & Honghai Liu. (2017). *"Hand gesture recognition based on convolution neural network"* published at Springer Science+Business Media, LLC, part of Springer Nature.

Ho-Joon Kim, Joseph S. Lee. & Jin-Hui Park. (2008). *"Dynamic Hand Gesture Recognition Using A Cnn Model With 3d Receptive Fields" IEEE Int. Conference Neural Networks & Signal Processing Zhenjiang*, China.

Hossain, M. S., Al-Hammadi, M. & Muhammad, G. (2019). "Automatic fruits Classification using deep learning for industrial applications," *IEEE Trans. Ind. Inform.*, vol. *15*, no. 2, pp. 1027–1034.

Muneer Al-Hammadi, Mansour Alsulaiman. & Ghulam Muhammad. (2019). *"Hand Gesture Recognition Using3D-CNN Model"* Published by the IEEE Consumer Electronics Society.

S. Kannadhasan and R, Nagarajan, Performance Improvement of H-Shaped Antenna With Zener Diode for Textile Applications, *The Journal of the Textile Institute*, Taylor & Francis Group, DOI: 10.1080/00405000.2021.1944523

S. Kannadhasan and R. Nagarajan, Development of an H-Shaped Antenna with FR4 for 1-10GHz Wireless Communications, *Textile Research Journal*, DOI:10.1177/00405175211003167 journals.sagepub.com/home/trj, March 21, 2021

S. Kannadhasan and R. Suresh, EMD Algorithm for Robust Image Watermarking. Recent Advances in Mechanical Engineering and Interdisciplinary Developments Organized by Ponjesly College of Engineering (ICRAMID 2014) Nagercoil on 7-8 March 2014, *Published for Advanced Materials Research* Vols.984-985 (2014) PP 1255-1260, ISSN No:1022-6680

S. Kannadhasan, G. Karthikeyan and V. Sethupathi, *A Graph Theory Based Energy Efficient Clustering Techniques in Wireless Sensor Networks. Information and Communication Technologies* Organized by Noorul Islam University (ICT 2013) Nagercoil on 11-12 April 2013, Published for Conference Proceedings by IEEE Explore Digital Library 978-1-4673-5758-6/13 @2013 IEEE.

S. Kannadhasan, M. Isavani and G. Karthikeyan, A Novel Approach Privacy Security Protocol Based SUPM Method in Near Field Communication Technology, *ArtificalArtificial Intelligence and Evolutionary Algorithms in Engineering Systems (ICAEES 2014)* Organized by Noorul Islam University, Kumaracoil Nagercoil on 23-24 April 2014 Published for Advances in Intelligent Systems and Computing

Shetty, S., Leung, T., Sukthankar, R. & Fei-Fei, L. (2014). "Large-scale video classification with convolutional neural networks," in *Proc. IEEE Conf. Comput. Vision Pattern Recognit.*, pp. 1725–1732.

Xiao Yan Wu. (2019). "*A hand gesture recognition algorithm based on DC-CNN*" published at Springer Science+Business Media, LLC, part of Springer Nature.

Xu, Z., Chen, X., Li, Y., Lantz, V., Wang, K. & Yang, J. (2011). "A framework for hand gesture recognition based on accelerometer and EMG sensors," *IEEE Trans. Syst.*, *Man*, *Cybern. Part A*, *Syst. Humans*, vol. *41*, no. 6, pp. 1064–1076.

In: Research Challenges in Science … ISBN: 978-1-68507-008-3
Editors: S. Kannadhasan et al.

Chapter 7

COMMUNITY ENERGY STORAGE: NATIONAL AND INTERNATIONAL SCENARIO - A REVIEW

Gauri M. Karve[1,*], Kalyani M. Kurundkar[2], and Geetanjali A. Vaidya[1], PhD

[1] Electrical Engg. Department, PVG's COET & GKPIM, Pune, Maharashtra, India

[2]Electrical Engg. Department, PVG's COET & GKPIM, Pune, Maharashtra, India

ABSTRACT

'Community Energy Storage (CES)' means a distributed group of small units of energy storage systems (ESS) connected to the load side of transformers which serves a small residential or small commercial load. CES is a storage owned electric utility which is distributed and located near to end-users.

The idea of CES is to dispatch and optimize energy storage resources. The potential benefits of dispensing the storage capacity rather than using one or a few bigger units can be significant which are backup power, flicker mitigation and integration of renewable energy sources (RES) at local level, while at grid level, benefits of CES are ancillary services, load leveling and power factor correction. The CES can be of various types such as pumped hydro, compressed air, flywheel or battery

* Corresponding Author's E-mail: gauri.m.karve@gmail.com.

storage. This work focuses on status and types of CES in India and in the world with their specific characteristics and capacities. The chapter is concluded with policy decisions regarding CES at few developed countries and its need in the other developing countries.

Keywords: community energy storage, national scenario, international scenario, review

INTRODUCTION

Community Energy Storage (CES) is distributed energy storage system (DESS), which is grid-connected and owned by the utility. DESSs are flexible storage systems which are positioned near to the end-users like- residential areas and businesses. The objective of CES concept is to project 'storage' as a source of energy for the future. Another unique feature of CES is that, it is similar to conventional utility alternatives and equipment [1]. The noteworthy advantage of using few smaller units than a bigger single unit is 'unit diversity.' Due to availability of so many storage units, it is hardly possible that CES power will be unavailable at any instant. This is very much important, when system reliability is considered. Due to the increase in grid connected solar systems, rooftop solar systems for houses and offices and rise in electric transportation, the load on local distribution system is increasing and hence the control of voltage and reliability of system is challenged more than earlier [2].

CES is an upcoming term for utility-owned that aggregate DESS which are positioned throughout a community, like water heaters, electric vehicles (EVs), and interconnected storage batteries, which are useful for increasing an efficacy of energy services to the customers [3]. The CES is intended to 'island', when a local distribution system is separated from the rest of the power grid electrically, CES can 'pick up' the consumer demand till stored energy is available. So, CES helps in providing ‘stand by’ power whenever needed in an emergency.

The functions of CES as:

1. It serves as a quick-responding and flexible option to generation.
2. CES also improves the reliability and power quality of local energy supply.
3. CES can be useful for offering the most of the types of 'ancillary services' such as frequency regulation, spinning reserve, etc., which are mandatory for maintaining the stability and reliability of the electrical grid.
4. It stores energy when the energy price is low and should use that energy when its price is high as an energy arbitrage.
5. In particular, CES is useful to maintain voltage stability at the distribution level.
6. As CES delivers power locally, so less transmission and distribution (T&D) losses and less requirement to fulfil the local 'peak demand.' Thus, CES can be helpful for 'load leveling' and 'peak shaving' as shown in Figure 1 and in Figure 2 [4].

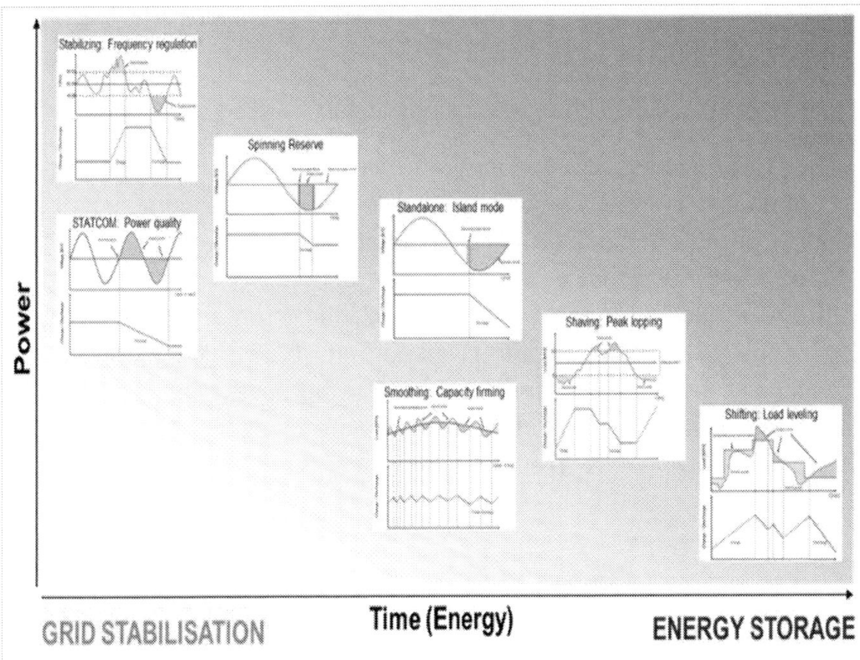

Figure 1. Functions of Energy Storage System [4].

Firstly, CES can be charged with wind or solar or from any other output from renewable generation. Like CES can be charged from wind energy at night after fulfilling the energy demand at that time and discharged in a day. While in case of solar energy, CES must be charged during a day after fulfilling the energy demand at that time and discharged at night. Hence CES can be helpful for ' leveling the load ' and ' shaving the peak demand' as shown in Figure 2 [5].

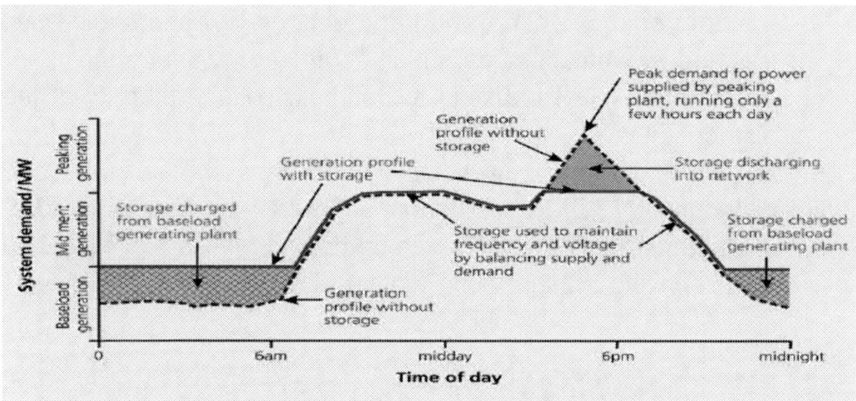

Figure 2. Effect of CESS on system demand-Load Leveling [5].

In some situations, if energy demand is less than the energy generated, then wind or solar generation has to be curtailed or the system operator has to pay someone to absorb that excess energy [5].

The demand for CES has grown as a result of the unpredictable nature of renewable energy sources and their integration into the grid. The energy storage systems used, in CES considering their applicability are majorly pumped hydro, compressed air, flywheel and various battery energy storage systems, refer Table 1 [6]. Table 1 shows few existing examples of CES all over the world.

Table 1. Various Energy Storage Systems used in CES [6]

ESS Tech. used in CES (With Capacity, Location)	Advantages	Disadvantages
Pumped Hydro Energy Storage (30 MW, Japan)	Oldest and Mature technology, Long cycle life (more than15000 cycles), largest power density, very small discharge losses	Slow response, needs more installation space and also dependent on geographical conditions, has an environment impact
Compressed Air Energy Storage (290MW Huntorf, Germany) and (110MW, Alabama, US) (115MW, US)	Independent on geographical conditions, High energy and power density, Long cycle life (more than 10000 cycles)	needs large installation space and also depends on geographical conditions, has an environment impact
Flywheel Energy Storage (20MW, US)	Temperature independent, environmentally friendly	Quick energy discharge, high cost of maintenance, needs good tensile strength of rotor material
Sodium Sulfur Battery Storage (NA-S) (34MW, Japan) and (300 MW, Japan) (2 MW, AEP), (4 MW, Texas)	High energy density (100-150 Wh/kg), Relatively high efficiency (70-80%), Long cycle life (more than 3000 cycles)	Relatively expensive (400-600 $/kWh), High temperature (300-350 oC), needs safety issues, available for large systems
Lithium-Ion Battery (Philadelhia, US)	High energy and power density (100-120 Wh/kg), High efficiency (85-95%)	Requires battery management to prevent overcharging, safety issues

COMMUNITY ENERGY STORAGE – INTERNATIONAL SCENARIO

The Community Energy Storage (CES) are majorly seen in Japan, United States and in Germany, which are discussed here. The U.S. Department of Energy (DOE) 'Grid 2030 Vision' Conference in April

2003, ESS turned to be a major concern in the future. In 2007, the DOE assembled an Electricity Advisory Committee (EAC) to make some guidelines for an energy road map for the United States (U.S.) including ESS. The EAC prepared a report providing guidelines for development and deployment of ESS in the U.S. grid for next ten years. Japan and Germany have been thinking so as to make large amount of ESS as an important part of their energy plan. Japan has targeted of 15% storage, while Germany and the U.S. planned for 10% and 2% of storage in the grid respectively [2].

The majority of the storage which is observed in any country is pumped hydro storage (PHS) which can be seen from Table 1, Table 2 and Table 3. These PHS plants are normally positioned at some distance away from load centers and has less impact on potential congestion issues that affect utility T&D networks during peak load demands.

In the world, the total available capacity of PHS plants is 95 GW, out of which roughly 22 GWs is operational in the U.S. and more 20 MW is upcoming [2].

The basic purpose of PHS plant was to support the off-peak or base-load demand. But now, PHS work has been extended to provide ancillary service like regulation of frequency because of the latest design of adjustable speed system, which is increasing the overall efficiency of the plant.

The first seawater pumped hydro power plant was built in Japan in the year 1999 (Yanbaru, 30 MW), which is roughly 3% of worldwide capacity of generation [5].

Earlier community energy storage (CES) was proven for PHS system and few systems with Compressed Air Energy Storage (CAES). One of the CAES plant is as shown in Figure 3 which is of 115 MW capacity, providing back up for 26 hours, located in the United States [2].

Figure 3. 115 MW, 26-Hr Compressed Air Energy Storage plant in the United States [2].

In 2005, CES concept was explored by American Electric Power (AEP) to appraise the advantages of positioning sodium sulfur (NaS) battery energy storage, rated at about 2 MW, at substations, which can give stand by power for 6 hours, as shown in Figure 4. The Figure 4 shows installation of NaS battery at a utility substation useful for managing peak load and improving supply reliability. In the United States, AEP installed three units of 2 MW, NaS battery systems, giving back up power for 7 hours. This system is as shown in Figure 4 which demonstrates that how intelligently ESS could be used in meeting requirements of smart grid [7].

Figure 4. Three, 2 MW, 6-hour NaS battery installation, American Electric Company (AEP) [7].

In 2009, DOE focused on using several types of distributed storage technologies in the range of 1 to 20 MW in addition to the 50 MWs which was already existing in the U. S. Hence in this line, eventually, AEP also thought off deploying small units of CES-BESS than one larger unit, So, instead of using one or two large battery systems with a rated power output of 2 MW at the distribution level substation, it is better to deploy 80 individual systems with 25kW power output as shown in Figure 5 and Figure 6 [2].

Figure 5. 25 kW, community energy storage system [2].

Source: American Electric Power (AEP).

Figure 6. Smaller units of CES than one larger unit, on secondary side of distribution transformer [2].

In addition to use of deployment of individual CES systems with 25kW power output, Figure 6 shows that special design concern is also taken care so that the CES could resemble with conventional utility equipment [2].

In 2009, the world's largest battery (34 MWs) came into service in support of uncertainty of a wind farm in Northern Japan. By 2010, the total worldwide installations of NaS battery are 365 MWs, out of which 300 MWs are operational in Japan [4].

In 1990, the development of the NaS battery system which is capable to deliver power for minimum 6 hours was demonstrated in Japan. While 1 MW NaS battery system was first time deployed at a utility substation in the US in 2006 [2].

Due to some vulnerable situations like long, very old and frequent outages transmission system, Texas deployed the largest 4 MW sodium-sulfur (NaS) battery in the United States, in Presidio in 2010. This battery system can respond quickly, can manage voltage fluctuations and momentary outages and provides back up power for eight hours. Thus, the battery power can be considered as a tool of self-healing distribution networks in case of transmission feeder loss.

After flywheel energy storage (FES) and supercapacitors, usage of sophisticated BESS technologies in DESS has been the primary focus due to the growing uses of battery systems in stationary and mobile applications. [8].

In 2009, the studies conducted showed that the efficacy of fast-response batteries and flywheels need roughly 40% less energy than that of fossil fuels for frequency regulation applications. And hence plants of BESS and FES are being built in the United States and other countries. The pilot systems for frequency regulation using battery systems and flywheel storage system are shown in Figure 7 and Figure 8 [9].

The experimental CES project using 1 MW capacity of lithium-ion batteries for managing ancillary service of frequency regulation is shown in Figure 7. This project is developed by PJM ISO and being in operation for last two years.

Figure 7. 1 MW, 15min frequency regulation pilot system at PJM With Li-Ion [9].

While Figure 8 shows pilot CES project with FES of 20 MW capacity used to regulate frequency and backing up the supply for 15 minutes.

Figure 8. 20 MW, 15-min flywheel frequency regulation plant [9].

Actually, CES is very powerful concept and provides use of a flexible, compact utility-owned storage solution for existing and upcoming issues which are faced by utilities in the current electricity market. It can be easily 'added' in the system whenever 'needed' as it is providing several benefits. Before utility engineers will agree to take storage as a standard substitute, it is imperative to standardize utility specifications for the CES systems. (AEP published 'operational specifications for CES' as an open source standard).

CES are essential for matching dynamic sources and loads at the distribution power grids. Like normal ESS, CES also can address renewable energy intermittency and integration, manage peak demands from electric vehicles and will stabilize microgrids.

The S&C Electric Company of Chicago installed BES/CES system to store off-peak power and re-dispatch it to the grid whenever required. Such storage systems being located near to the load centers have potential to improve the system's reliability, efficiency, capacity and can optimize two-way power flow and store RESs energy to fulfil peak demands [1].

There are several benefits of CES at different levels of power system grid. At the system level, CES provides capacity relief, frequency regulation, reactive power compensation and improvement in grid stability. While at the substation and feeder levels, CES provides peak shaving, load leveling, time-shifting of RES, voltage stabilization. And at the local level, CES guarantees reliable power through islanding and improves voltage control.

The Presidio, Texas, project is at substation level and provides power factor correction, by eliminating the need for fixed capacitor banks on distribution side. CES provides asset relief to the utilities by saving capital expenditures for major upgrades of substation and distribution systems defer [1].

As per the studies by the Electric Power Research Institute and Sandia National Laboratory, it is found that loads related to charging of electric vehicle (EV) and hybrid electric vehicle (HEV) will change rapidly. These load demand patterns will be quite unpredictable in the same geographical area and hence more challenging to address. In such situation, CES will manage these load demands very effectively and hence can reduce the stress on distribution transformers and power system during periods of peak demand.

Although CES is placed at the farthest end of the power grid, they are under the utility's administrative control, which can shave the peak or level the load of consumers so as to reduce overloads. Hence due to all these reasons, CES is being accepted increasingly as an effective smart tool for solving grid problems. As an example, S&C Electric Company of Chicago offers a radio-based system which controls groups of 1,000 CES units for managing load demand and other grid services.

CES is crucial for the future of power systems, which is already facing dynamic, large, unpredictable peak load demands and integration of variable renewable energy resources. It is a vital element for a reliable, efficient and responsive smart power system. The concept of CES has been established technically, now the next step is to upgrade utility regulations for the increasing deployment of grid-tied CES with reasonable compensation for the value it is providing.

The idea of the CES Initiative is to inspire manufacturers, utilities, and other market participants to discover and discuss their thoughts and experiences on CES opportunities. The response of stakeholders to CES Initiative is necessary until CES achieves its full potential to deliver a more flexible, commercial and environment-friendly electricity grid.

This section highlighted importance of CES, benefits of CES and status of CES at the international level.

Community Energy Storage - National Scenario

The status of CES at the international level is quite promising, but in India, still more focus on this issue is required. In India, less attention is given to CES due to some constraints, such as, most of the population stay in rural area and far away from electricity, so still in dark. Also, laying of cables, installation of transmission towers is tedious and uneconomical. But now, some promising things are going on such as foreign companies are interested to set up their storage plants in India. The two US-based companies as 'Imergy Power System' and 'Vionx Energy' want an India presence. Both companies are well-known in manufacturing 'flow batteries.' The flow batteries are advantageous as they are re-chargeable, can store adequate energy which will provide back-up for several hours and has life span of about 20 years. The Japanese company – 'Panasonic' wants to provide battery back-up to ATMs, in case of any power outage [10]. Another Chinese multinational company- 'Build Your Dreams (BYD) Co Limited' is interested in expansion mode in India through the 'Make in India' drive. The company is stepping up its rendezvous in India in the entire electric mobility chain and electronics with lithium-ion batteries and transforming it as a hub for the region. [11].

Now, India is on the edge of a 'storage revolution.' Recently IESA projected that Indian storage market will establish from 4.4 GW in 2015 to 70 GW/200 GWhr in 2022. Now a days, batteries are replacing diesel generators and maintaining grid stability alone, by capturing 4 to 6 GW of storage market.

Earlier the conventional lead-acid batteries were used for stationary and mobile applications due to their lower cost, but they have short life span (around 1,000 cycles), as compared to life span of lithium-ion batteries which is 3,000-5,000 cycles [10]. The use of lithium-ion batteries is limited because of their high cost. But the recent developments in batteries are so promising that their cost is expected to go down like cost of 'solar now' and cost of 'solar five years back.'

It is estimated that this is the time when large-scale storage systems initiated to be positioned in India. There are several ESS projects which are already existing and few are upcoming in India, are presented using Table 2 and Table 3. On the contrary to such developments, the AES system with 10 MWhr and Solar Energy Corporation of India (SECI)'s with 15 MWhr mark a great achievement.

According to the IEEFA report, some 2.6 GW of PHS is already operational in the country and another 3.2 GW of projects is under construction [12]. For the reference, the latest CES projects existing in India are pumped hydro storage (PHS) and battery energy storage (BES) systems which are shown in Figure 9 and Figure10 respectively.

The Figure 10 shows India's first grid-connected 10 MW battery-based community energy storage (CES) system. The project has been deployed at Tata Power Delhi Distribution Ltd's (Tata Power-DDL) sub-station in Rohini, Delhi in February 2019. This CES is offering several benefits like grid stability, enhancing reliability and protecting critical services for 2 million customers which are attended by the company [13].

Figure 9. 6.3 GW, Pumped Hydro Energy Storage plant in Andhra Pradesh in India [12].

Figure 10. 10 MW CES, installed in February 2019 in Delhi, India [13].

In India, though there are several ESS/CES projects which are existing and will come in near future, but there are few blockades to the proliferation of CES like technical challenges, initial investments, improving power availability scenarios, space constraints, environmental issues, net-metering policy, policy regulations and recommendations and last but not the least is the limitation of the business model adopted by developers.

Table 2. List of operational ESS projects in India [14]

Sr. No.	Name of the Storage Project/ Plant	Location of Project/ Plant	Rated Power of Project/ Plant (kW)
Pumped Hydro Storage Plant			
1	Sardar Sarovar Power Station	Navagam, Gujarat	14,50,000
2	Bhira Plant	Bhira, Maharashtra	1,50,000
3	Ghatghar Plant	Ghatghar, Maharashtra	2,50,000
4	Kadamparai Plant	Coimbatore, Tamil Nadu	4,00,000
5	Purulia Plant	Purulia, West Bengal	9,00,000
6	Tehri Plant	Tehri, Uttarakhand	10,00,000
7	Nagarjuna Sagar Project	Satrasala, Andhra Pradesh	7,00,000
8	Srisailam Plant	Srisailam, Andhra Pradesh	16,70,000
9	Kadana Plant	Kadana, Gujarat	2,40,000
10	Paithan (Jayakwadi Dam) Plant	Jayakwadi, Maharashtra	12,000
Solar Thermal Storage Plant			
11	India One Plant	Talheti, Rajasthan	1000
12	Clique Plant	Greater Noida, Uttar Pradesh	175
13	KVK Energy Solar Project	Askandra, Rajasthan	1,00,000
14	Gujarat Solar Plant	Kutch, Gujarat	25,000
15	Diwikar Molten Salt Thermal Storage	Askandra, Rajasthan	1,00,000
Electro-Chemical Storage			
16	Sun-carrier Omega Zero Energy Building (Vanadium flow battery)	Bhopal, Madhya Pradesh	45
17	Khareda Lakshmipura Microgrid (Lead Acid Battery)	Khareda Lakshmipura, Rajasthan	40
18	ACME BoB	Gurgaon, Haryana	250

Table 2. (Continued)

Sr. No.	Name of the Storage Project/ Plant	Location of Project/ Plant	Rated Power of Project/ Plant (kW)
19	GAT (Global Academy of Technology) College (Vanadium Flow Battery)	Bangalore, Karnataka	30
20	GigaCapacitor Hyderabad Test Project (IL)	Hyderabad, Andhra Pradesh	15,000

Table 3. Upcoming ESS projects in India [14]

Sr. No.	Project Name	Location	Rated Power (kW)
Solar Power Plant (CL-CSP) [Thermal Storage]			
1	Cross Linear Concentrated	Bhopal, Madhya Pradesh	30
Electro-Chemical Storage			
1	BESS – AES Haryana	Jhajjar, Haryana	10,000
2	Puducherry Smart Pilot Project	Puducherry	250
3	Pavagada Solar Park	Pavagada, Karnataka	5,000
4	L&T/e Camion-Anna University	Chennai, Tamil Nadu	125
5	Kadapa Ultra Mega Solar Park	Kadapa, Andhra Pradesh	5,000
6	500 kWh Battery Energy Storage System at CEL	Sahibabad, Uttar Pradesh	01
7	2.5 MW Solar Wind Hybrid Project with 1 MWh Energy Storage System	Rangreek, Himachal Pradesh	100
8	AEP-Mitsubishi-TATA Power (Li-ion)	Rohini, New Delhi	1,00,000

To understand this issue, technical challenges faced by the Indian battery market are such as - battery charging pattern, memory effect, surge current, high discharge rate, C-rate, maintenance, deep-discharge batteries, battery management system, operational challenges, high depth of discharge, distilled water requirement, distilled water refill frequency, acid spillage (especially for lead acid batteries), transportation and replacement; while commercial challenges as - capital cost, high maintenance cost, advance battery technologies, high dependence on battery manufacturer, custom duties and after sale services etc. Irrespective of these technical barriers, lack of policy decisions become a main hurdle for implementation of CES/BESS systems on large scale in India.

CONCLUSION

The paper states local and global advantages of Community Energy Storage (CES) system as backup power, flicker mitigation, integration of RES, peak load shaving at substation, power factor correction and ancillary services. The paper describes various types of CES as Pumped Hydro Energy Storage, Compressed Air Energy Storage, Flywheel Energy Storage and Battery Energy Storage with their capacity size and operational status in all over the world.

The paper highlights that only few developed countries like Germany, US and Japan have policies for implementation of CES/ESS systems which made it possible to deploy large number of CES projects in those countries.

However other developing countries like India needs strong policy decisions regarding CES so as to implement it on a large-scale considering benefit to all stake holders like utility, storage owners, prosumers and consumers.

The paper concluded with technical and economic challenges faced by the Indian CES market.

REFERENCES

[1] ESA - Energy Storage Association - *Community Energy Storage - Executive Summary,* NW, Washington, DC info@energystorage. org.

[2] Bradford P. Roberts, Life Senior Member IEEE, and Chet Sandberg, Fellow IEEE. The Role of Energy Storage in Development of Smart Grids, *Proceedings of IEEE*, *Vol. 99,* No.6, June 2011, pp. 1139-1144.

[3] Robin Roy, NRDC, *Community Energy Storage for Cleaner, Lower-Cost Electricity*, April 25, 2016.

[4] Integrating High Levels of Renewables into microgrid: Opportunities, Challenges, Strategies, *A GTM Research White Paper*, sponsored by ABB, February 2016.

[5] Ami Joseph & M. Shahidepour, ECE Dept, Illinois Institute of Technology, Chicago, Illinois, USA Battery Energy Storage Systems in Electric Power Systems; *IEEE* 2006, pp. 1-8.

[6] Haoran Zhao, Qiuwei Wu. Review of energy storage system for wind power integration support. In *Applied Energy;* ELSEVIER 2014, pp. 1-9.

[7] *IEEE Power Energy Magazine, Bottling Energy Issue*, *vol. 7*, no. 4, Jul./Aug. 2009.

[8] Multan Biswas, Md Shafiul Azim, Tonmoy Kumar Saha, Umama Zobayer, Monalisa Chowdhury Urmi. Towards Implementation of Smart Grid: *An Updated Review on Electrical Energy Storage Systems Md 2013*, pp. 122-132.

[9] D. Rastler, B New demand for energy storage, *Edison Electr. nst., Electr. Perspect., Sep./Oct. 2008*, pp. 30–47.

[10] *Energy Today, for Smart Energy Decision Makers, 'Energy News',* 19th August 2016 - Article by Dr. Rahul Walawalkar. (www.energytoday.co.in).

[11] V Rishi Kumar, "THE HINDU," Economy, Business Line, *'With Make IN India, BYD Looks to Make Country EV Hub For The region'*, Published on 17 December, 2019.

[12] Anand Gupta, "Andhra Pradesh Invite Bids for Detailed Report to Set Up 6.3 GW of Pumped Storage Hydro Projects," *IEEFA report*, September 10, 2020.

[13] THE HINDU, Economy, Business Line, *'Grid-scale energy storage system set up in Delhi'*, Published on February 13, 2019. www.tatapower-ddl.com.

[14] Amit Kumar Rohit, Saroj Rangnekar, "An overview of energy storage and its importance in Indian renewable energy sector:

Part II – energy storage applications, benefits and market potential", *Journal of Energy Storage*, 13, 13 July 2017, pp. 447–456.

About the Authors

Gauri M. Karve, ME Electrical [Power Systems], Assistant Professor in Electrical Engg. Dept. PVG's COET & GKPIM, Pune. Areas of interest: Electrical Power System, Renewable energy, Energy Storage. E-mail: gauri.m.karve@gmail.com.

Kalyani M. Kurundkar, ME Electrical (Power Systems), Assistant Professor in Electrical Engg. Dept. PVG's COET & GKPIM, Pune. Areas of interest: Renewable energy, Energy Market, Ancillary Services. E-mail: kalyani.kurundkar@gmail.com

Geetanjali A Vaidya, ME [Electrical Power systems], PhD. Electrical Engineering, Professor of Electrical Engg. Dept. PVG's COET & GKPIM, Pune, Member IEEE, Areas of interest: Sustainable energy, Energy efficiency, Electricity pricing, Electric vehicle. E-mail: geetvaidya@gmail.com.

In: Research Challenges in Science ... ISBN: 978-1-68507-008-3
Editors: S. Kannadhasan et al.

Chapter 8

MODELLING OF MULTI-INPUT THREE LEVEL DC-DC CONVERTER

K. K. Saravanan * ***and B. Rajeshkanna***
University College of Engineering, Thirukkuvalai, India

ABSTRACT

A multi-input DC-DC boost converter is proposed to obtain power from several input sources and to supply the regulated output voltage to the load from the sources. The structure of Multiple-Input Converter (MIC) is simpler compared to the use of several single-input converters for each source. As the conventional source of energy is rapidly depleting and the cost of energy is rising.in the multi-input converters offer a cost effective solution in applications which requires multiple input sources such as fuel cell vehicles and renewable energy systems. The basic idea is to integrate a number of converters in either input dc/dc conversion stage or in isolation stage, in addition to commonly shared output stage. There are various multi-input topologies proposed in literature based on non-isolate and isolated structures. In, a non-isolated multi-input buck/boost converter sharing the same switch has been proposed. For multi-input converters provide galvanic isolation, majority of the work has been devoted to phase-shifted full-bridge converters. In, two identical current fed

* Corresponding Author's E-mail: saravanan.santi@gmail.com.

full-bridge converters are connected in parallel to interface a common output rectifying diode-bridge. Similar concept has been applied using individual output rectification stages but sharing an integrated transformer in two current fed bridges are controlled in a phase-shifted manner. The parallel connections of multi-input cells are first identified as pulsating voltage-source cell and pulsating current-source cell, and various topological combinations have been presented.

Keywords: full-bridge rectifier, Dc/Dc Converter, pulsating current cell

INTRODUCTION

The greater efficiency and tighter control features of power electronics are becoming attractive for applications in motion control by replacing the earlier electromechanical and electronic systems [1-10].Applications in power transmission and renewable energy include high-voltage dc (VHDC) converter stations, flexible ac transmission system (FACTS), static var compensators, and energy storage[13]. In power distribution, these include dc-to-ac conversion, dynamic filters, frequency conversion, and custom power system. Almost all new electrical or electromechanical equipment's, from household air conditioners and computer power supplies to industrial motor controls, contain power electronic circuits and or systems [18-21]. In order to keep up, working engineers involved in control and conversion of power and energy into applications ranging from several hundred voltages at a fraction of an ampere for display devices to about 10,000 V at high-voltage dc transmission should have a working knowledge of power electronics [14-15].

PROPOSED MULTI-INPUT ISOLATED THREE-LEVEL DC/DC CONVERTER

Topology Description

The proposed multi-input converter is essentially based on the operation of the three-level isolated dc/dc converter. The secondary side rectifier can be either a half-bridge rectifier accompanied by a tapped winding for low output voltage application, or a full-bridge rectifier for high output voltage applications as shown in Figure 1.

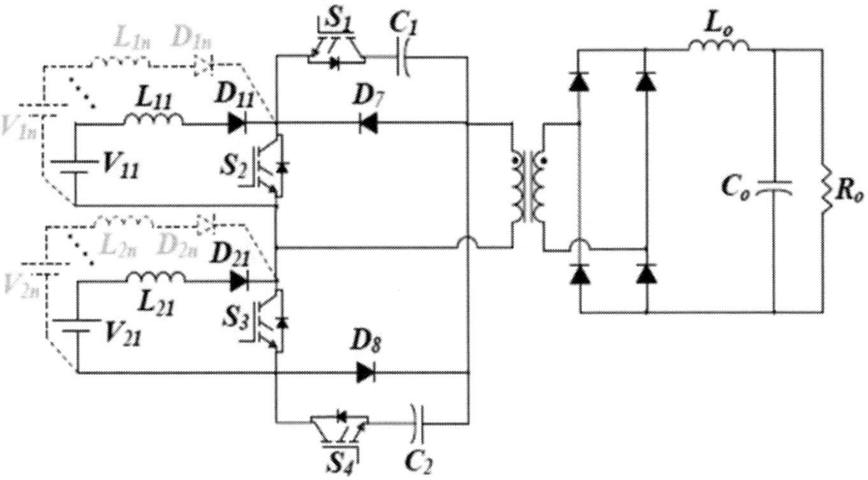

Figure 1. Proposed multi input isolated three-level DC/DC converter.

Since the PWM scheme here is similar to that in three-level converters, the control complexity is not intensive. The boost inductors, connected to the input sources of different voltages, are charged when $S_2$and $S_3$are turned on, respectively. When the related switch (S_2or S_3) is turned off, the energy stored in the inductor is transferred to the load. The excess or insufficient energy is either absorbed or compensated by the dc link capacitors. At the same time, $S_1$to $S_4$are switched to apply $V_{dc}/2$, $-V_{dc}/2$, and zero voltage across the primary side of the transformer. In comparison with the two-stage counterpart, two active switches and boost diodes are eliminated, while two blocking diodes are added to block the reverse current from the dc link capacitors. The switching scheme of the converter is given in Figure 2.

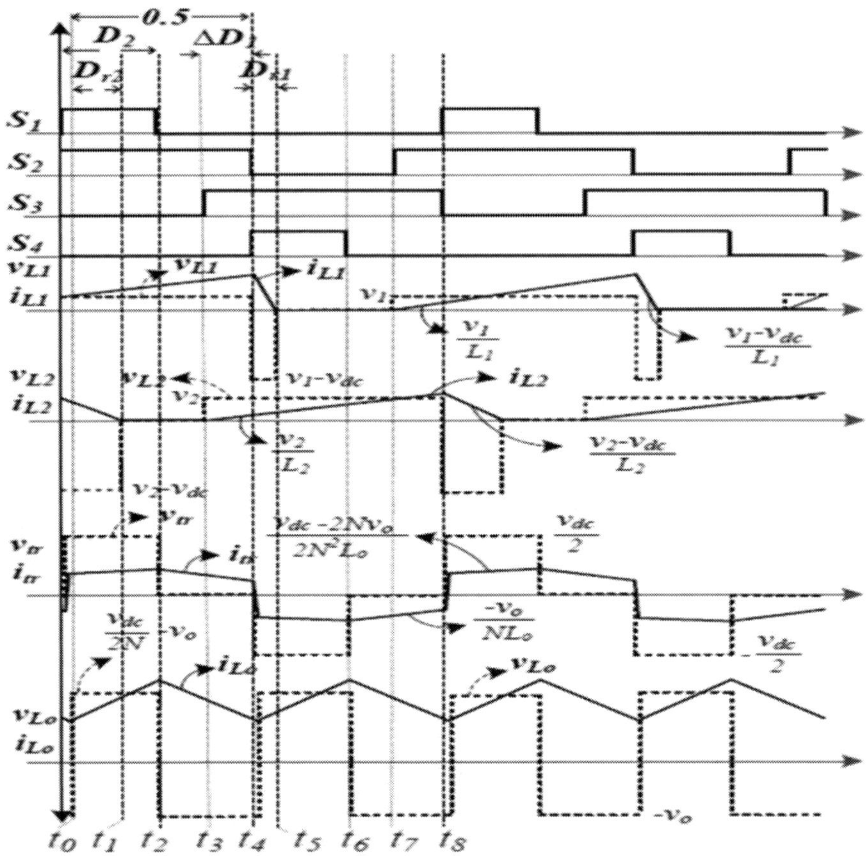

Figure 2. Gating scheme and key waveforms.

The switches S2-S3 and S1-S4 have 180° phase shifts with respect to each other. The duty cycle of the middle switches should be greater than 50% such that they allow a freewheeling path for the transformer primary side current. The switching scheme is as follows; S1 is turned on right after S3 is turned off, and similarly, S4 is turned on when S2 is turned off. A dead-time should be inserted in between the turning on the instant of S1 and turning off instant of S3, and likewise between switching of S2 and S4 to avoid short circuits.

Operation Principle

The operation modes of the circuits, which are given in this section. It should be noted that these operation modes are valid when iL2>iL1. For this case, iL2, pk is higher than the primary side of the transformer at a steady state. Under this condition, iL1, pk is smaller than the primary side of the transformer at steady-state. If iL1>iL2, the equivalent circuits would be different from the one presented here.

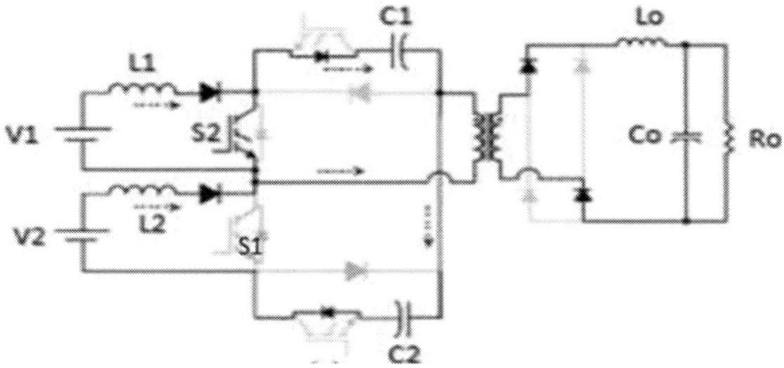

Figure 3.

Interval 1 [t<t0]: Before t0, S3 is off and S1 is on. The energy stored in L2 is released to the load through the primary winding of the transformer. The excess energy is transferred to C1 by conducting the body diode of S1. Thus, S1 can be turned on under zero voltage with proper time delay. The current of L1 increases linearly under V1. The total current of L2 charges C2. Meanwhile, –Vdc/2 is applied to the primary side of the transformer.

$$d\begin{bmatrix} i_{L1}(t) \\ i_{L2}(t) \\ V_{C1}(t) \\ V_{C2}(t) \\ i_{Lo}(t) \\ V_{Co}(t) \end{bmatrix} / dt = \begin{bmatrix} V_1 / L_1 \\ \left(V_2 - V_{C1}(t) - V_{C2}(t)\right) / L_2 \\ i_{L2}(t)/C_1 - N i_{Lo}(t)/C_1 \\ i_{L2}(t)/C_2 \\ V_{C1}(t)/NL_o - V_{Co}(t)/L_o \\ i_{Lo}(t)/C_o - V_{Co}(t)/C_o R_o \end{bmatrix} \quad (1)$$

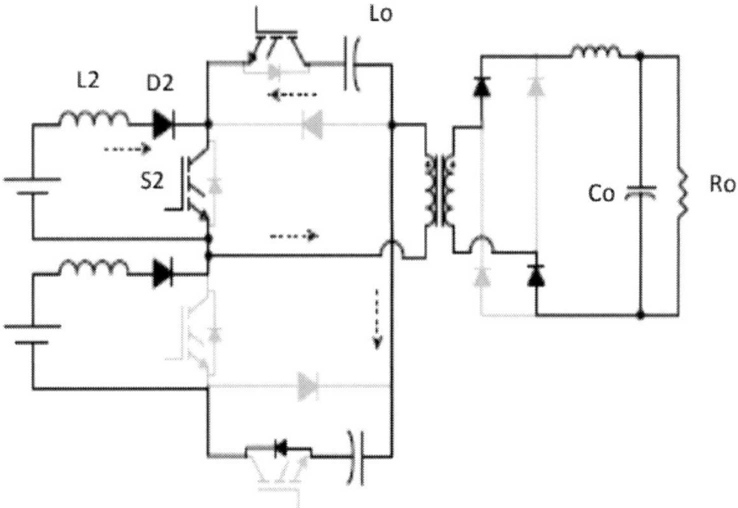

Figure 4.

Interval 2 [t0<t<t1]: At t=t0, the current of L1 becomes equal to that of the primary side of the transformer. From this moment on, C1 discharges over the primary side of the transformer to the load. The current of C2 is equal to that of L2.

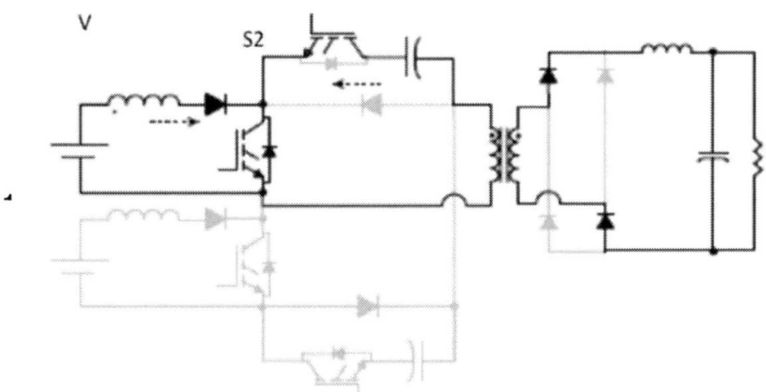

Figure 5.

Interval 3[$t_1<t<t_2$]:At $t=t_1$, the currents of L_2 and C_2 reach to zero. The load current is solely supplied by C_1. The current of L_1 continues to store energy under the influence of V1.

$$d\begin{bmatrix} i_{L1}(t) \\ i_{L2}(t) \\ V_{C1}(t) \\ V_{C2}(t) \\ i_{Lo}(t) \\ V_{Co}(t) \end{bmatrix} / dt = \begin{bmatrix} V_1/L_1 \\ 0 \\ -Ni_{Lo}(t)/C_1 \\ 0 \\ V_{C1}(t)/NL_o - V_{Co}(t)/L_o \\ i_{Lo}(t)/C_o - V_{Co}(t)/C_oR_o \end{bmatrix} \tag{2}$$

Interval 4 [t2<t<t3]: At the beginning of this interval, S1 is turned off. The current in the leakage inductance conducts D7 and the primary side current freewheels, hence, zero voltage is applied across the primary side of the transformer. The output inductor voltage is equal to –Vo and the output inductor current decreases linearly.

$$d\begin{bmatrix} i_{L1}(t) \\ i_{L2}(t) \\ V_{C1}(t) \\ V_{C2}(t) \\ i_{L0}(t) \\ V_{C0}(t) \end{bmatrix} / dt = \begin{bmatrix} V_1/L_1 \\ 0 \\ 0 \\ 0 \\ -V_{C0}(t)/L_0 \\ i_{L0}(t)/C_0 - V_{C0}(t)/C_0R_0 \end{bmatrix} \tag{3}$$

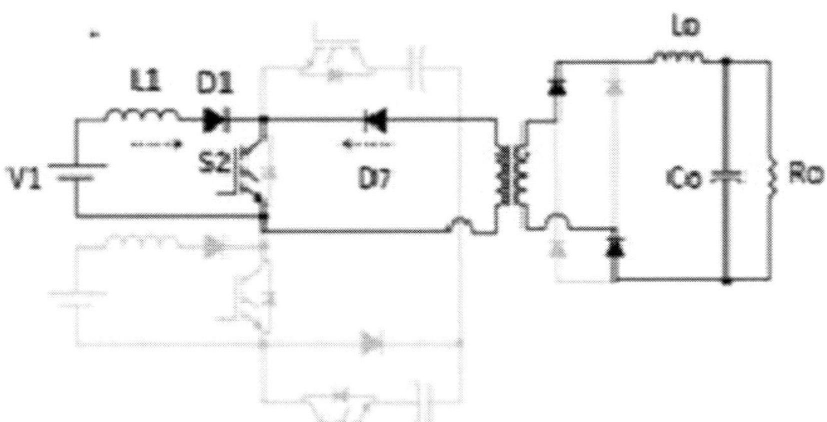

Figure 6.

Interval 5 [t3<t<t4]: At t=t3, S3 is turned on, while S2 is kept on. The primary side current continues to freewheel and zero voltage is applied across the primary side; hence, the output inductor current continuous to decrease under output voltage. Meantime, V2 is applied across L2, and current increases linearly storing energy in L2.

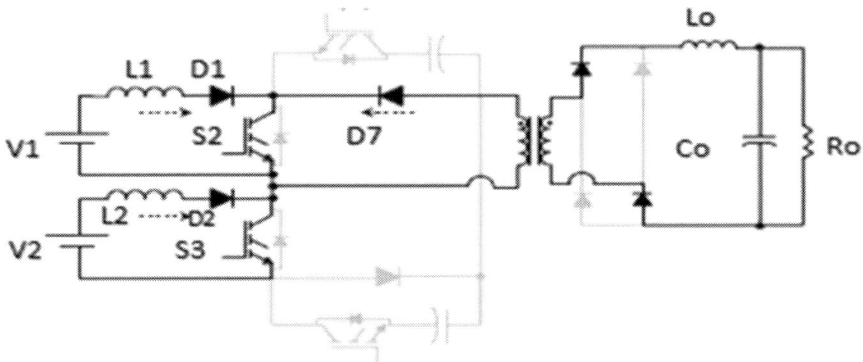

Figure 7.

$$d\begin{bmatrix} i_{L1}(t) \\ i_{L2}(t) \\ V_{C1}(t) \\ V_{C2}(t) \\ i_{L0}(t) \\ V_{C0}(t) \end{bmatrix} / dt = \begin{bmatrix} V_1 / L_1 \\ V_2 / L_2 \\ 0 \\ 0 \\ -V_{C0}(t)/L_0 \\ i_{L0}(t)/C_0 - V_{C0}(t)/C_0R_0 \end{bmatrix} \quad (4)$$

Interval 6 [$t_4<t<t_5$]: At $t=t_4$, S_2 is turned off. The stored energy in L_1 is released to the load as well as to C_1. As stated before, peak current of L_1 is smaller than the load current for this case. Thus, C_2 discharges over to the load and $V_{dc}/2$ is applied across the primary side of the transformer. The current of L_2 continues to store energy under the influence of V_2.

$$d\begin{bmatrix} i_{L1}(t) \\ i_{L2}(t) \\ v_{C1}(t) \\ v_{C2}(t) \\ i_{Lo}(t) \\ v_{Co}(t) \end{bmatrix} / dt = \begin{bmatrix} (V_1 - v_{C1}(t) - v_{C2}(t)) / L_1 \\ V_2 / L_2 \\ i_{L1}(t) / C_1 \\ i_{L1}(t) / C_2 - N i_{Lo}(t) / C_2 \\ v_{C2}(t) / N L_o - v_{Co}(t) / L_o \\ i_{Lo}(t) / C_o - v_{Co}(t) / C_o R_o \end{bmatrix} \tag{5}$$

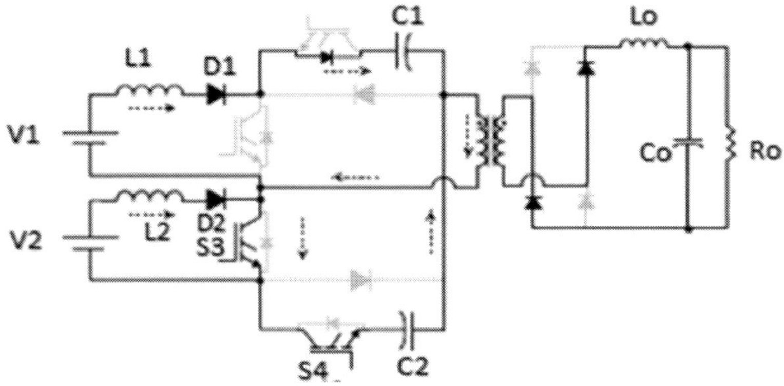

Figure 8.

Interval 7 [$t_5<t<t_6$]: At $t=t_5$, the energy stored in L_1 is completely transferred out. Its current reaches to zero at DCM. The load current is solely supplied by C_2.

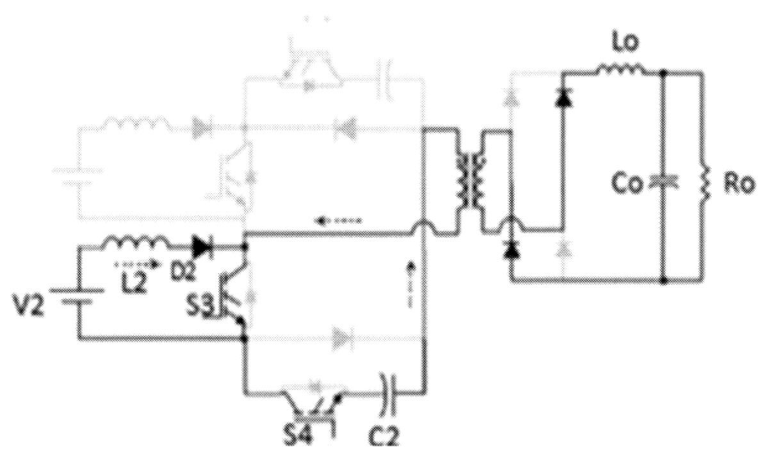

Figure 9.

$$d\begin{bmatrix} i_{L1}(t) \\ i_{L2}(t) \\ v_{C1}(t) \\ v_{C2}(t) \\ i_{Lo}(t) \\ v_{Co}(t) \end{bmatrix} / dt = \begin{bmatrix} 0 \\ V_2 / L_2 \\ 0 \\ -Ni_{Lo}(t) / C_2 \\ v_{C2}(t) / NL_o - v_{Co}(t) / L_o \\ i_{Lo}(t) / C_o - v_{Co}(t) / C_o R_o \end{bmatrix} \quad (6)$$

Interval 8 [$t_6<t<t_7$]: In this interval, S_4 is turned off. The current in the leakage inductance conducts D_8 and the primary side current freewheels, hence, zero voltage is applied across the primary side of the transformer. The current of L_2 continues to increase.

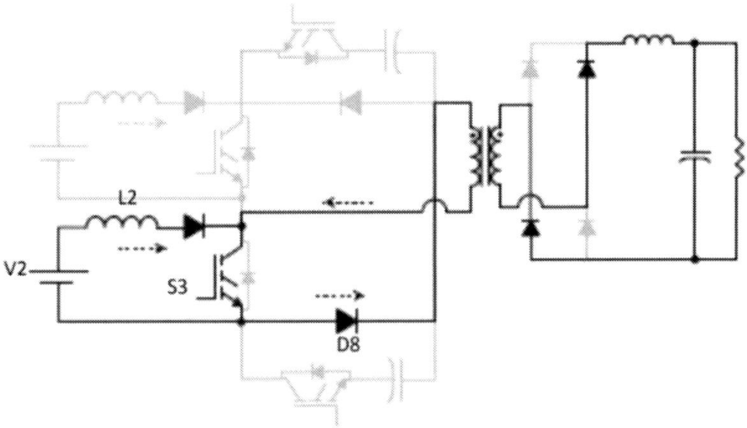

Figure 10.

$$d\begin{bmatrix} i_{L1}(t) \\ i_{L2}(t) \\ v_{C1}(t) \\ v_{C2}(t) \\ i_{Lo}(t) \\ v_{Co}(t) \end{bmatrix} / dt = \begin{bmatrix} 0 \\ V_2 / L_2 \\ 0 \\ 0 \\ -v_{Co}(t) / L_o \\ i_{Lo}(t) / C_o - v_{Co}(t) / C_o R_o \end{bmatrix} \quad (7)$$

Interval 9 [$t_7<t<t_8$]: At $t=t_7$, S_2 is turned on while S_3 remains at on-state. The primary side current continues to circulate through D_8. The current of both inductors L_1 and L_2 increase linearly under the applied input voltages

$$d\begin{bmatrix} i_{L1}(t) \\ i_{L2}(t) \\ v_{C1}(t) \\ v_{C2}(t) \\ i_{Lo}(t) \\ v_{Co}(t) \end{bmatrix} / dt = \begin{bmatrix} V_1 / L_1 \\ V_2 / L_2 \\ 0 \\ 0 \\ -v_{Co}(t) / L_o \\ i_{Lo}(t) / C_o - v_{Co}(t) / C_o R_o \end{bmatrix} \tag{8}$$

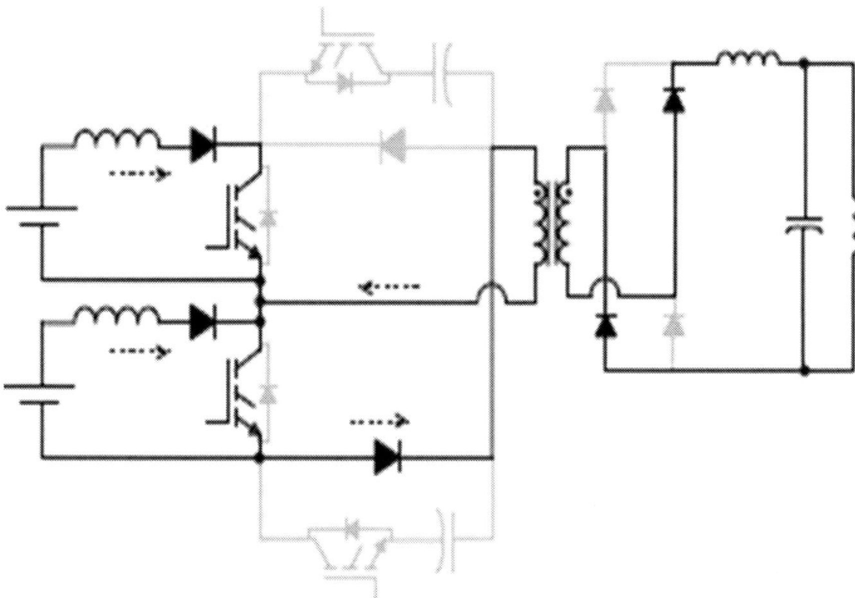

Figure 11.

SIMULINK MODEL

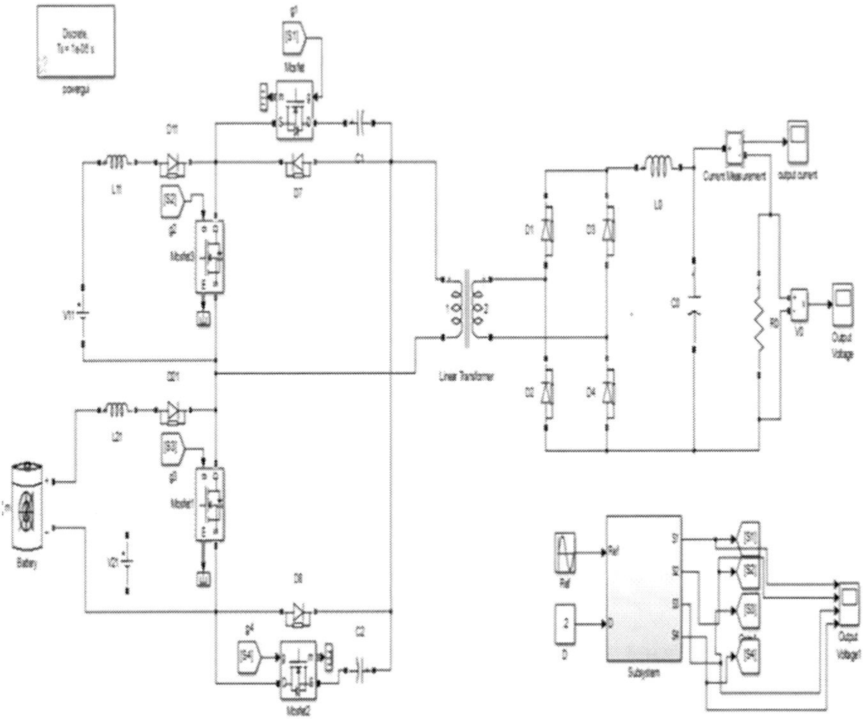

Figure 12. Simulink Model of Multi-input DC-DC Converter.

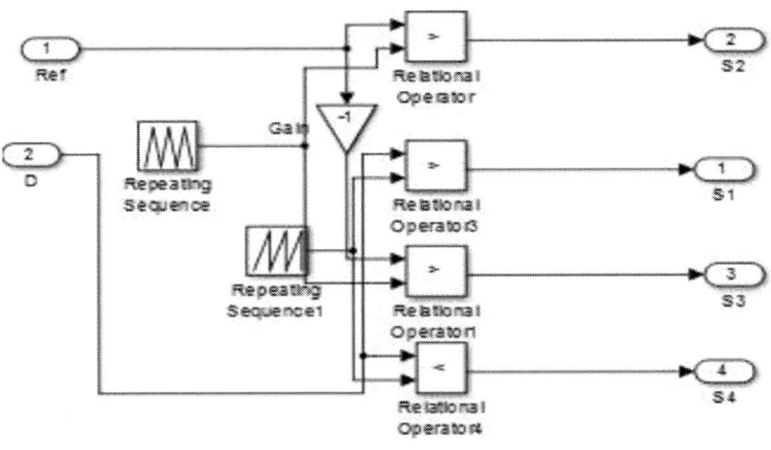

Figure 13. Subsystem of multi-input dc/dc converter.

Simulation Output

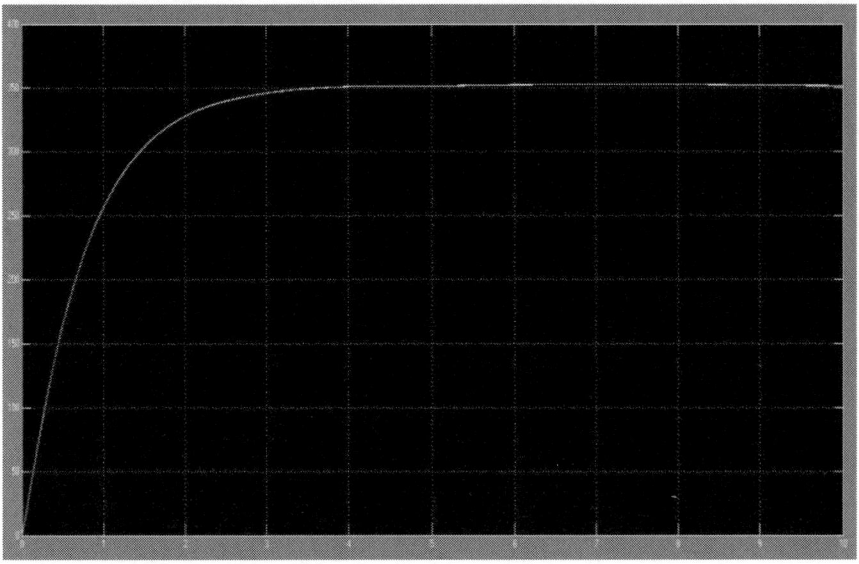

Figure 14. Output of DC voltage waveform.

The voltages retain at 365.87 V and the input voltage is given as 210 V through dc battery and voltage input sources.

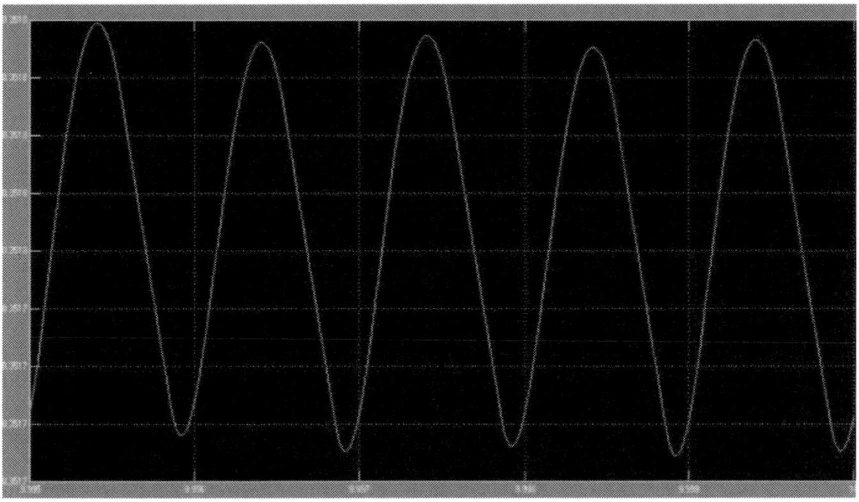

Figure 15. Output of DC current waveform.

Conclusion

In this chapter, a new multi-input dc/dc converter having the same number of dynamic switches starting three-level isolated dc/dc converter without presenting extra switching actions is proposed. The circuit examination and configuration thought have been given in detail. With the appropriate determination of input inductors, autonomous load sharing can be accomplished. To confirm the operation of the converter simulation has been performed. The outcomes demonstrate the effective integrated operation of input sources with a three-level disconnected dc/dc structure, through working four switches in a phase-shifted manner without presenting control complexity.

References

[1] Liu F., Z. Wang, Y. Mao and X. Ruan, "Asymmetrical Half-Bridge Double-Input DC/DC Converters Adopting Pulsating Voltage Source Cells for Low Power Applications," *IEEE Trans. Power Electron.*, vol. 29, no. 9, pp. 4741-4751, Sept. 2014.

[2] Napoli A. D., F. Crescimbini and S. Rodo, "Multiple input dc/dc converter for fuel-cell powered hybrid vehicles," in *Proc. IEEE Power Electron. Spec. Conf. (PESC)*, 2002, pp. 1685–1690.

[3] Chiu H. J., H. M. Huang, L. W. Lin and M. H. Tseng, "A multiple-input dc/dc converter for renewable energy system," in *Proc. IEEE Int. Conf. Ind. Technol.*, 2005, pp. 1304–1308.

[4] Chen Y. M., Y. C. Liu and T. F. Wu, "Multi-input dc–dc converter based on the multiwinding transformer for renewable energy applications," *IEEE Trans. Ind. Appl.*, vol. 38, no. 4, pp. 1096–1104, Jul./Aug. 2002.

[5] Duarte J. L., M. Hendrix and M. G. Simoes, "Three-port bidirectional converter for hybrid fuel cell systems," *IEEE Trans. Power Electron.*, vol. 22, no. 2, pp. 480–487, Mar. 2007.

[6] Danyali S., S.H. Hosseini and G. B. Gharehpetian, "New extendable single stage multi-input DC–DC/AC boost converter," *IEEE Trans. Power Electron.*, vol. 29, no. 2, pp. 775–788, Feb. 2014.

[7] Dobbs B. G. and P. L. Chapman, "A Multiple-Input DC-DC Converter Topology," *IEEE Power Electron. Lett.*, vol. 1, no. 1, pp. 6-9, Mar. 2003.

[8] Khaligh A., A. M. Rahimi, Y-J. Lee, J. Cao, A. Emadi, S. D. Andrews, C. Robinson and C. Finnerty, "Digital Control of an Isolated Active Hybrid Fuel Cell/Li-Ion Battery Power Supply, *IEEE Trans. Veh. Technol.*, vol. 56, no. 6, pp. 3709-3721, Nov. 2007.

[9] Kobayashi K., H. Matsuo and Y. Sekine, "Novel Solar-Cell Power Supply System Using a Multiple-Input DC–DC Converter," *IEEE Trans. Ind. Electron.*, vol. 50, no. 1, pp. 281-286, Feb. 2006.

[10] Onar O. C. and A. Khaligh "A Novel Integrated Magnetic Structure Based DC/DC Converter for Hybrid Battery/ Ultracapacitor Energy Storage Systems," *IEEE Trans. Smart Grid*, vol. 3, no. 1, pp. 296-307, Mar. 2012.

[11] Marchesoni M. and C. Vacca, "New DC–DC Converter for Energy Storage System Interfacing in Fuel Cell Hybrid Electric Vehicles," *IEEE Trans. Power Electron.*, vol. 22, no. 1, pp. 301-286, Jan. 2007.

[12] Liu F., Z. Wang, Y. Mao and X. Ruan, "Asymmetrical Half-Bridge Double-Input DC/DC Converters Adopting Pulsating Voltage Source Cells for Low Power Applications," *IEEE Trans. Power Electron.*, vol. 29, no. 9, pp. 4741-4751, Sept. 2014.

[13] Marchesoni M. and C. Vacca, "New DC–DC Converter for Energy Storage System Interfacing in Fuel Cell Hybrid Electric Vehicles," *IEEE Trans. Power Electron.*, vol. 22, no. 1, pp. 301-286, Jan. 2007.

[14] Khaligh A., J. Cao and Y.-J. Lee, "A Multiple-Input DC–DC Converter Topology," *IEEE Trans. Power Electron.*, vol. 24, no. 3, pp. 862-868, Mar. 2009.

[15] S. Kannadhasan and R, Nagarajan, Performance Improvement of H-Shaped Antenna With Zener Diode for Textile Applications, *The Journal of the Textile Institute*, Taylor & Francis Group, DOI: 10.1080/00405000.2021.1944523

[16] Wai R.-J., C.-Y. Lin, J.-J. Liaw and Y.-R. Chang, "Newly Designed ZVS Multi-Input Converter," *IEEE Trans. Ind. Electron.*, vol. 58, no. 2, pp. 555-566, Feb. 2011.

[17] Li Y., X. Ruan, D. Yang, F. Liu and C. K. Tse, "Synthesis of Multiple-Input DC/DC Converters," *IEEE Trans. Power Electron.*, vol. 25, no. 9, pp. 2372-2385, Sept. 2010.

[18] Liu D. and H. Li, "A ZVS Bi-Directional DC–DC Converter for Multiple Energy Storage Elements," *IEEE Trans. Power Electron.*, vol. 21, no. 5, pp. 1513-1517, Sept. 2006.

[19] Wang Q., J. Zhang, X. Ruan and K. Jin, "Isolated Single Primary Winding Multiple-Input Converters," *IEEE Trans. Power Electron.*, vol. 26, no. 12, pp. 3435-3442, Dec. 2011.

[20] S. Kannadhasan and R. Nagarajan, Development of an H-Shaped Antenna with FR4 for 1-10GHz Wireless Communications, *Textile Research Journal*, DOI:10.1177/00405175211003167, 21 March 2021

[21] Dusmez S., X. Li and B. Akin, "A Fully Integrated Three-Level Isolated Single-Stage PFC Converter, *IEEE Trans. Power Electron.*, vol. 30, no. 4, pp. 2050-2062, Apr. 2015.

[22] Kwasinski A., "Identification of Feasible Topologies for Multiple-Input DC–DC Converters," *IEEE Trans. Power Electron.*, vol. 24, no. 3, pp. 856-861, Mar. 2009.

[23] Yuan-mao Y. and K. W. E. Cheng, "Multi-input voltage-summation converter based on switched-capacitor," *IET Power Electron.*, vol. 6, no. 9, pp. 1909-1916, Nov. 2013.

In: Research Challenges in Science … ISBN: 978-1-68507-008-3
Editors: S. Kannadhasan et al.

Chapter 9

E-LEARNING AND BEYOND: THE FUTURE OF TECHNOLOGY IN EDUCATION

***K. V. Binoy Mathew*[1,*]**
***and Maryelizabeth Tidiya Walarine*[2]**
[1]Department of Physiotherapy,
Composite Regional Centre for Skill Development,
Rehabilitation and Empowerment of Persons with Disabilities,
Kozhikode, Kerala, India
[2]Department of Medical-Surgical Nursing, MIMS College of Nursing,
Malappuram, Kerala, India

ABSTRACT

The era of knowledge explosion forces us to be lifelong learners to be abreast in our chosen field. Advancements in information and communication technology have revolutionized technology-enhanced learning. The dawn of the twenty-first century has opened to us the vast world of online learning. Online learning methods have taken the place of traditional learning

* Corresponding Author's E-mail: binoy9797@gmail.com.

methods during the pandemic. Many online platforms have already put their footprints in the current educational scenario. Multidimensional learning methods are blended together to give a much better outcome than single-dimensional learning methods. Innovative learning methods have won the hearts of students. They have the potentiality to assist in achieving the comprehensive development of a student by bringing them out into the realm of practicability. The flipped classroom, Escape rooms, Gamification, M-learning, Immersive Virtual Reality, Competency-based education, Digital Problem Based Learning are some examples of such innovative learning methods.

Keywords: Active Learning, Blended Learning, E-learning, Flipped Learning, M-Learning, Online Learning

Introduction

We are passing through the era of knowledge explosion. Millions of information and learning contents are getting added daily to the internet-based knowledge databases and social media platforms. To excel in the chosen field, the students, as well as professionals, need to upgrade their knowledge and skills. This forces one to be lifelong learners.

There are currently many resources available to assist in the journey of lifelong learning. Advancements in information and computer technology along with the widespread adoption of digital equipment had caused a surge in online and digital-based learning programs. COVID-19 Pandemic which spread worldwide from December 2019 forced mass shifting of educational activities from the traditional classroom setting to online mode.

E-Learning

E-learning as well as online learning brought in the facility of temporal and geographical flexibility. E-learning or electronically supported learning can be online or offline. Online learning involves the use of electronic technology and internet-based media for delivering, supporting, and enhancing learning and teaching. Online learning is increasingly adopted in health education. Massive Open Online Courses (MOOC), 3 D virtual learning, educational smartphone applications, and web-based platforms are being extensively used (McCutcheon et al.2018).

The first MOOC was conducted by the University of Manitoba. MOOC allows enrollment of numerous students. They are mostly delivered free of cost or with nominal fees. The courses are with well-defined learning objectives, quality multimedia learning content, assessments, and certification. The courses are delivered through internet-based platforms (Mahajan et al. 2019). The government of India is conducting hundreds of MOOCs through its SWAYAM portal. The students have the opportunity to transfer the credits earned through MOOCs to their academic records through this portal (Mahajan et al. 2019). MOOCs are increasingly being used by health professionals as a channel for continuing professional development, enhancing research skills, and capacity building.

Online Learning Strategies for Active Learning

Active learning includes any instructional method where the student actively engages in the learning process. The active learning approach helps students to gain deeper knowledge by promoting self-directed learning. This helps them to upgrade themselves by being lifelong learners (Coyne et al. 2018). Active learning involving meaningful and interesting activities is shown to influence learners' professional awareness (Hummel et al. 2021). Narrative Animated Videos (NAVs) with immersive story-based contents are believed to promote self-directed and problem-based learning (Hull et al. 2019). During the pandemic time, students preferred recorded video lectures over live classes, as it provided the freedom to self-pace the study time and duration as per their convenience (Camargo et al. 2020).

Program community for online degree learners was found to be of great relevance. Social media, synchronous communications, and face-to-face interactions are used by faculty to foster community activities.

Lack of time, resources, interest on the part of learners and faculty as well as poor reward policies from the organizations were considered as barriers to their successful functioning (Bolliger et al. 2019).

Social media has been widely used for knowledge sharing and dissemination in recent years. Facebook, Twitter, WhatsApp, Instagram, and Pinterest have been increasingly used for sharing multimedia learning content (Ahmed et al. 2020). A Multi-center randomized controlled trial cautioned the use of WhatsApp for learning purposes. The results of the study showed that the use of WhatsApp leads to worse clinical reasoning in comparison to traditional e-learning (Clavier et al. 2019).

Blended learning involves the incorporation of Information and Communication Technology (ICT) applications along with traditional face-to-face instructions (Bennett et al. 2020). Student intrinsic motivation and active engagement are integral for successive learning. The right choice and blend of technologies in the learning process play a key role in the process. The desire to combine the benefits of face-to-face learning and online learning led to the development of blended learning. A randomized trial comparing blended and online learning on clinical supervising skills in nursing students showed that blended learning improved students' motivation, attitudes, satisfaction, and subject-specific knowledge (McCutcheon et al. 2018). A review showed that blended learning with video-based online resources is a helpful tool to enhance clinical skills among health professional students (Coyne et al. 2018). A study reported significant improvements in student's motivation, satisfaction, and mood state after blended learning with m-learning but could not produce many changes in educational outcomes in comparison to traditional training (Lozano-Lozano et al. 2020). Contrary to it, a meta-analysis found that blended learning leads to better knowledge outcomes when compared to traditional learning in health education (Vallée et al. 2020).

The flipped classroom, a type of blended learning, is widely used nowadays. Flipped classrooms equip learners to apply the acquired knowledge in solving the problems in their chosen area and come up with effective solutions. Here the teachers act as a guide in the solution-finding process (Sailer and Sailer 2021). The students are provided with self-study materials that are rich in multimedia content so that they can self-pace their initial learning activities. Later students are engaged in application-based training based on the contents learned (Sailer and Sailer 2021). A review showed that health professional students were quite happy with flipped classroom-based training. They were specifically satisfied with accessible pre-class video contents and also group-based discussions in a classroom setting (Ramnanan and Pound 2017). A meta-analysis (Hew and Lo 2018) comparing flipped classrooms to traditional face-to-face instruction showed an overall significant effect in favor of the flipped classroom approach for health professions education (SMD = 0.33, 95% CI 0.21–0.46, $p < 0.001$). It was found that flipped classroom approach improved practical skills and saved time in surgery courses (Dombrowski et al. 2018). It was supported by a meta-analysis that showed that flipped learning was more effective than lecture-based learning in improving nursing student's theoretical scores (Li et al. 2020).

Escape rooms are live team-based games where they have to complete a task within a specific period of time. Besides providing entertainment, they are now increasingly used as a learning tool (Veldkamp et al. 2020). Escape rooms foster active learning, enhance teamwork, collaboration, and communication skills. Reported challenges for establishing educational escape rooms include a restricted budget, inability to set up it in the classroom, and lack of time (Veldkamp et al. 2020).

Gamification in education has shown a positive effect on learner performance and motivation(Bai et al. 2020; Sailer and Homner 2020). Gamified quiz platforms like Kahoot have been increasingly used in educational settings, which provide instant feedback. Immediate task level feedbacks like this in terms of points or scores provide great motivation.

This results in excellent learning performance and later application of knowledge(Sailer and Sailer 2021).A meta-analysis(Bai et al. 2020) on effectiveness of gamification in student learning performance in various educational settings showed a significant medium effect (Hedges' g = 0.504, 95%CI [0.284–0.723], $p < 0.001$).The reasons for learners' enjoyment in gamification were the fact that it fosters enthusiasm, provides feedback on performance, promotes goal-setting, and fulfills the need for recognition(Bai et al. 2020). Research on the effect of gamification on students in entrepreneurship courses showed that it resulted in enhanced student engagement, experience, and entrepreneurial self-efficacy (Isabelle 2020).

M-learning or mobile learning or Smartphone-based learning gained wide popularity in recent years. It is the learning approach where educational contents are accessed at anytime from anywhere through mobile or smartphone (Lozano-Lozano et al. 2020). A meta-analysis showed that m-learning is equal to or better than traditional face-to-face learning (Dunleavy et al. 2019).

Immersive Virtual Reality (IVR) gives the perception of presence in a non-world environment. IVR has become affordable and accessible in recent times, and it is being increasingly used in educational settings for learning purposes (Bower and Jong 2020). It has high potential in educational settings as it makes learning more engaging and motivating (Freina and Ott 2015). Virtual patients are interactive computer simulations of clinical scenarios used for health professional education. Virtual patient simulated learning has shown the potential in improving clinical reasoning skills, procedural skills, and team-based skills in health professional education (Kononowicz et al. 2019).

Competency-based education has been adopted worldwide in recent years. In this educational approach, the learners engage in learning activities and acquire knowledge and skills at their personalized pace with guidance from faculty to achieve the well-defined program outcomes. The learners are provided credentials after the completion and evaluation of the assignments (Boyer and Bucklew 2019). It has been widely accepted that credit hours are an insufficient measurement of learning (Garn 2019).

The learning pedagogy in Asian countries like China and India, needs to shift from traditional credit-based learning outcomes to more learner-driven competency-based ones. The faculty needs to be trained as effective facilitators.

Digital Problem Based Learning (DPBL) has been found to be effective as Problem Based Learning (PBL) and superior to traditional learning in improving knowledge and skills (Tudor Car et al. 2019).DPBL may be fully digital or in the form of blended learning. Problem-based learning has been used in education and specifically in health professional education to enhance decision-making, problem-solving and critical thinking skills. In PBL, learning is student-centered. Small collaborative groups are utilized for learning purposes and teachers act as guides.PBL involves 3 parts; Presentation and analysis of the problem, self-directed learning, solution synthesis, and reporting (Yew and Goh 2016).

Facilitating workplace learning opportunities is mutually beneficial for employers as well as employees. Human capital is one of the vital assets of any successful organization. One of the world's leading innovators, International Business Machines (IBM) has developed its own educational programs, and employees are provided digital badges after successful completion of programs. Besides, they collaborate with Northeastern University with an option for conversion of these industrial credits earned for academic degree credit (Leaser et al. 2020). The factors found to facilitate the successful completion of online courses were interactive sessions, opportunities for peer collaboration, locally adapted content, a self-directed pace, multi-modality options, and organizational support (Archer et al. 2020).

Employers prefer employees who had completed degrees done through face-to-face instruction rather than online mode (Roberto and Johnson 2019). Online-based learning will flourish only if there happens a drastic shift in such employer preferences favoring new modes of learning.

Recommendations

- Live online courses/synchronous video contents lead to learner fatigue. So it is suggested to occasionally turn the video camera off or look away from the screen(Ebner and Greenberg 2020).
- Develop learning pedagogy and online courses based on neuro-cognitive principles facilitating binge-watching (Ebner and Greenberg 2020).
- The teachers should be provided organizational support and training in online instruction and multimedia learning content development.

- The online program should have the ideal amount of synchronous and asynchronous educational elements, to reap the benefits of both.

References

Ahmed, S., Zimba, O.and Gasparyan, A. Y. (2020). Moving towards online rheumatology education in the era of COVID-19. *Clinical Rheumatology*, 39(11), 3215 - 3222. https://doi.org/10.1007/s10067-020-05405-9.

Archer, A., Berry, I., Bajwa, U., Kalda, R.and Di Ruggiero, E. (2020). Preferred modalities for delivering continuing education to the public health workforce: a scoping review. *Health Promotion and Chronic Disease Prevention in Canada*, 40(4), 116 - 125. https://doi.org/10.24095/hpcdp.40.4.03.

Bai, S., Hew, K. F.and Huang, B. (2020). Does gamification improve student learning outcome? Evidence from a meta-analysis and synthesis of qualitative data in educational contexts. *Educational Research Review*, 30, 100322. https://doi.org/10.1016/j.edurev.2020.100322.

Bennett, D., Knight, E.and Rowley, J. (2020). The role of hybrid learning spaces in enhancing higher education students' employability. *British Journal of Educational Technology*, 51(4), 1188 - 1202. https://doi.org/10.1111/bjet.12931.

Bolliger, D. U., Shepherd, C. E.and Bryant, H. V. (2019). Faculty members' perceptions of online program community and their efforts to sustain it. *British Journal of Educational Technology*, 50(6), 3283 - 3299. https://doi.org/10.1111/bjet.12734.

Bower, M.and Jong, M. S. (2020). Immersive virtual reality in education. *British Journal of Educational Technology*, 51(6), 1981–1990. https://doi.org/10.1111/bjet.13038.

Boyer, N.and Bucklew, K. (2019). Competency-based education and higher education enterprise systems. *The Journal of Competency-Based Education*, 4(1), e01180. https://doi.org/10.1002/cbe2.1180.

Camargo, C. P., Tempski, P. Z., Busnardo, F. F., de Arruda Martins, M.and Gemperli, R. (2020). Online learning and COVID-19: A meta-synthesis analysis. *Clinics*, 75, e2286. https://doi.org/10.6061/clinics/2020/e2286.

Clavier, T., Ramen, J., Dureuil, B., Veber, B., Hanouz, J.-L., Dupont, H., Lebuffe, G., Besnier, E. and Compere, V. (2019). Use of the Smartphone App WhatsApp as an E-Learning Method for Medical Residents: Multicenter Controlled Randomized Trial. *JMIR MHealth and UHealth*, 7(4), e12825. https://doi.org/10.2196/12825.

Coyne, E., Rands, H., Frommolt, V., Kain, V., Plugge, M.and Mitchell, M. (2018). Investigation of blended learning video resources to teach health students clinical skills: An integrative review. *Nurse Education Today*, 63, 101 - 107. https://doi.org/10.1016/j.nedt.2018.01.021.

Dombrowski, T., Wrobel, C., Dazert, S.and Volkenstein, S. (2018). Flipped classroom frameworks improve efficacy in undergraduate practical courses – a quasi-randomized pilot study in otorhinolaryngology. *BMC Medical Education*, 18(1), 294. https://doi.org/10.1186/s12909-018-1398-5.

Dunleavy, G., Nikolaou, C. K., Nifakos, S., Atun, R., Law, G. C. Y.and Tudor Car, L. (2019). Mobile Digital Education for Health Professions: Systematic Review and Meta-Analysis by the Digital Health Education Collaboration. *Journal of Medical Internet Research*, 21(2), e12937. https://doi.org/10.2196/12937.

Ebner, N.and Greenberg, E. E. (2020). Designing Binge-Worthy Courses: Pandemic Pleasures and COVID-19 Consequences. *Negotiation Journal*, 36(4), 535 - 560. https://doi.org/10.1111/nejo.12339.

Freina, L.and Ott, M. (2015). A literature review on immersive virtual reality in education: State of the art and perspectives. *Proceedings of ELearning and Software for Education, (ELSE)* (Bucharest, Romania, April, 23 - 24, 2015), 8.

Garn, M. (2019). Three decades and still counting (the wrong things): An analysis of three reports on the possibility and practicality of shifting our academic currency from credit hours to competencies. *The Journal of Competency-Based Education*, 4(3), 1 - 9. https://doi.org/10.1002/cbe2.1196.

Hew, K. F.and Lo, C. K. (2018). Flipped classroom improves student learning in health professions education: a meta-analysis. *BMC*

Medical Education, 18(1), 38. https://doi.org/10.1186/s12909-018-1144-z.

Hull, D. M., Lowry, P. B., Gaskin, J. E.and Mirkovski, K. (2019). A storyteller's guide to problem-based learning for information systems management education. *Information Systems Journal*, 29(5), 1040 - 1057. https://doi.org/10.1111/isj.12234.

Hummel, H. G. K., Nadolski, R. J., Eshuis, J.and Slootmaker, A. (2021). *Serious Game in Introductory Psychology for Professional awareness : Optimal learner control and authenticity*, 52(1), 125 - 141. https://doi.org/10.1111/bjet.12960.

Isabelle, D. A. (2020). Gamification of Entrepreneurship Education. *Decision Sciences Journal of Innovative Education*, 18(2), 203 - 223. https://doi.org/10.1111/dsji.12203.

Kononowicz, A. A., Woodham, L. A., Edelbring, S., Stathakarou, N., Davies, D., Saxena, N., Tudor Car, L., Carlstedt-Duke, J., Car, J.and Zary, N. (2019). Virtual Patient Simulations in Health Professions Education: Systematic Review and Meta-Analysis by the Digital Health Education Collaboration. *Journal of Medical Internet Research*, 21(7), e14676. https://doi.org/10.2196/14676.

Leaser, D., Jona, K.and Gallagher, S. (2020). Connecting Workplace Learning and Academic Credentials via Digital Badges. *New Directions for Community Colleges*, 2020(189), 39 - 51. https://doi.org/10.1002/cc.20396.

Li, B.-Z., Cao, N.-W., Ren, C.-X., Chu, X.-J., Zhou, H.-Y.and Guo, B. (2020). Flipped classroom improves nursing students' theoretical learning in China: A meta-analysis. *PLOS ONE*, 15(8), e0237926. https://doi.org/10.1371/journal.pone.0237926.

Lozano-Lozano, M., Fernández-Lao, C., Cantarero-Villanueva, I., Noguerol, I., Álvarez-Salvago, F., Cruz-Fernández, M., Arroyo-Morales, M.and Galiano-Castillo, N. (2020). A Blended Learning System to Improve Motivation, Mood State, and Satisfaction in Undergraduate Students: Randomized Controlled Trial. *Journal of Medical Internet Research*, 22(5), e17101. https://doi.org/10.2196/17101.

Mahajan, R., Gupta, P.and Singh, T. (2019). Massive Open Online Courses: Concept and Implications. *Indian Pediatrics*, 56(6), 489 - 495. https://doi.org/10.1007/s13312-019-1575-6.

McCutcheon, K., O'Halloran, P.and Lohan, M. (2018). Online learning versus blended learning of clinical supervisee skills with pre-registration nursing students: A randomised controlled trial. *International Journal of Nursing Studies*, 82, 30 - 39. https://doi.org/10.1016/j.ijnurstu.2018.02.005.

Ramnanan, C.and Pound, L. (2017). Advances in medical education and practice: Student perceptions of the flipped classroom. *Advances in Medical Education and Practice*, Volume 8, 63 - 73. https://doi.org/10.2147/AMEP.S109037.

Roberto, K. J.and Johnson, A. F. (2019). Employer Perceptions of Online Versus Face-to-Face Degree Programs. *Journal of Employment Counseling*, 56(4), 180 - 189. https://doi.org/10.1002/joec.12132.

S. Kannadhasan and R, Nagarajan, Performance Improvement of H-Shaped Antenna With Zener Diode for Textile Applications, *The Journal of the Textile Institute, Taylor & Francis Group*, DOI: 10.1080/00405000.2021.1944523

S. Kannadhasan, M. Isavani and G. Karthikeyan, A Novel Approach Privacy Security Protocol Based SUPM Method in Near Field Communication Technology, *ArtificalArtificial Intelligence and Evolutionary Algorithms in Engineering Systems* (ICAEES 2014) Organized by Noorul Islam University, Kumaracoil Nagercoil on 23-24 April 2014 Published for Advances in Intelligent Systems and Computing

S. Kannadhasan and R,Nagarajan, Performance Improvement of H-Shaped Antenna With Zener Diode for Textile Applications, *The Journal of the Textile Institute, Taylor & Francis Group*, DOI: 10.1080/00405000.2021.1944523

Sailer, M.and Homner, L. (2020). The Gamification of Learning: A Meta-analysis. *Educational Psychology Review*, 32(1), 77 - 112. https://doi.org/10.1007/s10648-019-09498-w.

Sailer, M.and Sailer, M. (2021). *Gamification of in-class activities in flipped classroom lectures,52*(1), 75 - 90. https://doi.org/10.1111/bjet.12948.

Tudor Car, L., Kyaw, B. M., Dunleavy, G., Smart, N. A., Semwal, M., Rotgans, J. I., Low-Beer, N.and Campbell, J. (2019). Digital Problem-Based Learning in Health Professions: Systematic Review and Meta-Analysis by the Digital Health Education Collaboration. *Journal of Medical Internet Research*, 21(2), e12945. https://doi.org/10.2196/12945.

Vallée, A., Blacher, J., Cariou, A.and Sorbets, E. (2020). Blended Learning Compared to Traditional Learning in Medical Education: Systematic Review and Meta-Analysis. *Journal of Medical Internet Research*, 22(8), e16504. https://doi.org/10.2196/16504.

Veldkamp, A., Daemen, J., Teekens, S., Koelewijn, S., Knippels, M. P. J.and Joolingen, W. R. (2020). Escape boxes: Bringing escape room experience into the classroom. *British Journal of Educational Technology*, 51(4), 1220 - 1239. https://doi.org/10.1111/bjet.12935.

Yew, E. H. J.and Goh, K. (2016). Problem-Based Learning: An Overview of its Process and Impact on Learning. *Health Professions Education*, 2(2), 75 - 79. https://doi.org/10.1016/j.hpe.2016.01.004.

In: Research Challenges in Science … ISBN: 978-1-68507-008-3
Editors: S. Kannadhasan et al.

Chapter 10

STATE OF CHARGE ESTIMATION TECHNIQUES USING THE FILTER-BASED AND DATA-DRIVEN APPROACH: A REVIEW

Siddhi Vinayak Pandey[1,*] and Soham Shah[2]

[1]Department of Electrical Engineering, Adani Institute of Infrastructure Engineering, Ahmedabad, Gujarat, India

[2]Department of Information & Communication Technology, Adani Institute of Infrastructure Engineering, Ahmedabad, Gujarat, India

ABSTRACT

The demand of batteries in electrical systems is increasing significantly. Many electrical vehicles (automobiles), various power plants and space satellites are using different types of batteries based on their applications. Among these, the quantification of State of Charge (SOC) is the consequential parameters for all the battery packs. The SOC represents available battery power in authentic time domain, which gives the information to users about how much time it can provide a potency to the loads. This paper describes various methods to estimate SOC in order to predict

* Corresponding Author's E-mail: siddhipandey.ele17@aii.ac.in.

remaining battery energy in real time domain. However, there has been significant research on the SOC estimation techniques, out of which the most broadly used and highly accurate methods are described in this paper. The filter-based technologies such as - Kalman Filter (KF), Extended Kalman Filter (EKF), Unscented Kalman Filter (UKF)/Sigma Point Kalman Filter (SPKF) and various data driven techniques as - Support Vector Machines (SVMs), various Neural Networks such as - NARX NNs, NIO NNs, RNNs and DNNs are being discussed in detail. The Reinforcement Learning approach; which is one of the widely used Hybrid-estimation method for the accurate estimation of SOC has also been deliberated. The paper gives the literature survey and elucidates these technologies in considerable detail and culminates the future research requires in this field along with many different efficient estimation techniques for several types of linear and nonlinear battery loads.

Introduction

Due to less availability of fossil fuels, natural gas, and nuclear energy [1]; Human beings are ceaselessly endeavoring to shift towards renewable and green energy-based resources, which causes very less pollution as compared to non-renewable based resources [2]. In field of electrical technologies, all the equipment, contrivances and machines need electrical power supply to perform its functionality, which requires enormous capacity of energy storage contrivances. Among them, batteries are one of the pioneer energy storage contrivances used in electrical systems. The working range of the batteries are rigorously predicated on its energy density, charging discharging life cycles, no load discharge rate and temperature range. According to demand of diverse applications, different types of batteries are being used. The life cycle and efficiency of batteries are directly based on accurate measurement of its SOC. However, it is essential to know the accurate SOC value to predict the running time upto which it can provide a power to the load. The direct measurement techniques were primitively used that included methods to measure Open Circuit Voltage (OCV), terminal voltage, impedance, etc. The Coulomb counting approach is a simple approach to measure the battery SOC. Though, Coulomb Counting method has certain limitations and is comparatively less accurate with respect to other estimators. So, researchers are shifting towards discovering the accurate techniques like different model-based, filter-based and data-driven methods for the prediction of accurate SoC in real time domain.

The filter-based techniques like Kalman Filters (KF), extended Kalman Filters (EKF), Unscented Kalman Filter (UKF), etc. are some of the widely used approaches for the estimation of SOC in real time domain. In recent years, the data-driven approach is being used to gain higher accuracy. Techniques such as - SVM (Support Vector Machines), Radial Bias Function NNs (RBFNNs), Fuzzy NNs (FNNs), Artificial Neural networks (ANNs) were used initially [3].

With the recent advancements in machine learning, RBF-based SVMs, Non-linear Auto-regressive with Exogenous input Neural Networks (NARXNNs), NIO NNs, Recurrent Neural Networks (RNNs) and Deep Neural Networks (DNNs) are some of the most efficient and commonly used data-driven techniques. Reinforcement Learning (RL) approach is a hybrid estimation technique, which has recently started developing for estimating the battery SOC.

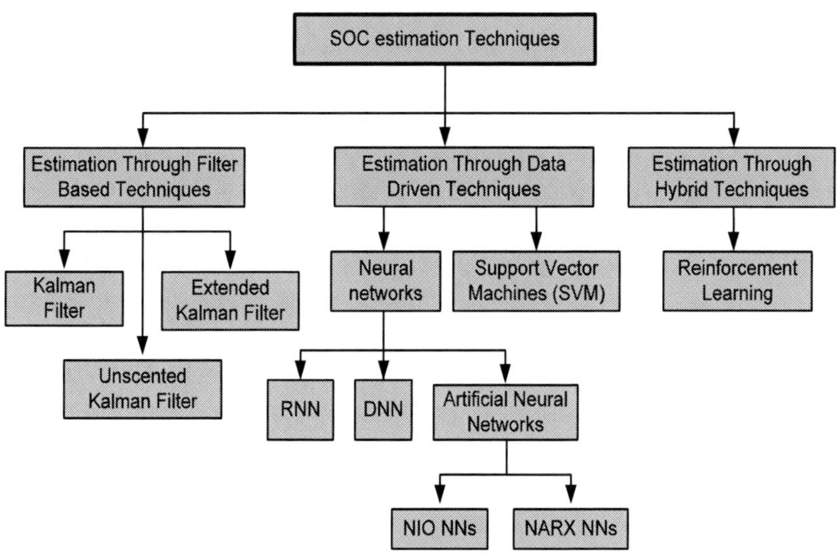

Figure 1. SOC Estimation Techniques.

In this paper, detailed description about the various filter based and data driven State of Charge estimation approaches is provided along with the required model architecture, mathematical equations and concept, figures, experimental results and graphs for better understanding of the concept. The above attached Figure 1 is the types of modern SOC estimation techniques, which is being used for better prediction of available battery power in authentic time domain.

FILTER BASED ESTIMATION METHODS FOR SOC

In filter-based estimation method, different types of statistical techniques [4] such as Bayesian theorems, Taylor series, Matrices & Determinants and various mathematical analogies are used to determine the accurate estimation of SOC with nonlinear battery models [5]. Some of the famous model-based estimation methods for SOC detection techniques are illustrated as below-

SOC Estimation Using Kalman Filter

The accurate measurement of SOC is a difficult task in real time domain. Because the straightforward measurement method is not applicable for estimation of SOC in live condition. Kalman Filter can be used to overcome this challenge [6]. As mentioned in Zhang 2011 [7] and He 2013 [8] - with the combination of Coulomb Counting technique, Kalman filter estimates the accurate overvoltage and current in real time domain.

Kalman Filter takes an input from battery in terms of - current, voltage and temperature. After receiving all the data, it quickly estimates the outcome of the target parameter that is under measurement. Many times, the readings contain uncertain variation of voltage and current corresponding to time [9].

The linear Kalman Filter state space model predict the current state (k) from the previous state (k-1). Based on the current state K, the %SOC of battery is calculated. In Figure 2 [10], the simple state-space model is explained using Kalman filter. By putting the values of voltage, current and temperature in state space matrix, Kalman gain (Kk) is calculated for the %SOC estimation in real time domain.

SOC Estimation Using Extended Kalman Filter

The KF is designed for linear discrete time dynamical system while the EKF is applied for discrete time nonlinear system [11]. Due to increase in temperature, the nonlinearity of OCV increases which affects the battery SOC [12]. To overcome this, EKF is used which automatically reduces the error of the system by just measuring the terminal voltage and current of a battery.

Yuan et al. 2013 [13] - used the Auto Regressive Exogenous (ARX) battery model and EKF for accurate SOC estimation of a Lithium-ion battery, which was suitable for the real time applications. The graphical representation of ARX battery model with EKF is mentioned in the Figure 3 [13].

After conducting this experiment, Yuan et al. 2013 [13] observed that - the EKF decreases the extreme terminal voltage error from 0.0397 to 0.0165 V. While the mean error value drops from 0.0014 to 0.0007 V. This type of improvement can increase the efficiency of batteries in real time domain.

Linear – state space model:

$$x_{k+1} = A_k x_k + B_k u_k + w_k$$
$$y_k = C_k x_k + D_k u_k + v_k$$

Initialization, For, k=0, set -

$$\hat{x}_0^+ = E[x_0]$$
$$P_{\hat{x},0}^+ = E[(x_0 - x_0^+)(x_0 - x_0^+)^T]$$

Computation,

For, k=1,2,3......... compute

State estimate time update: $\hat{x}_0^- = A_{k-1}\hat{x}_{k-1}^- + B_{k-1}u_{k-1}$

Error covariance time update: $P_{\hat{x},0}^- = A_{k-1}P_{\hat{x},0}^+ A_{k-1}^T + Q_w$

Kalman Gain matrix: $K_k = P_{\hat{x},k}^- C_k^T [C_k P_{\hat{x},0}^- C_k^T + R_v]^T$

State estimate measurement update: $\hat{x}_k^+ = \hat{x}_k^- + K_k[y_k - C_k\hat{x}_k^- - D_k u_k]$

Error covariance measurement update: $P_{\hat{x},k}^+ = (I - k_k C_k)\, P_{\hat{x},k}^-$

Figure 2. [10] Linear State Space Model of Kalman Filter.

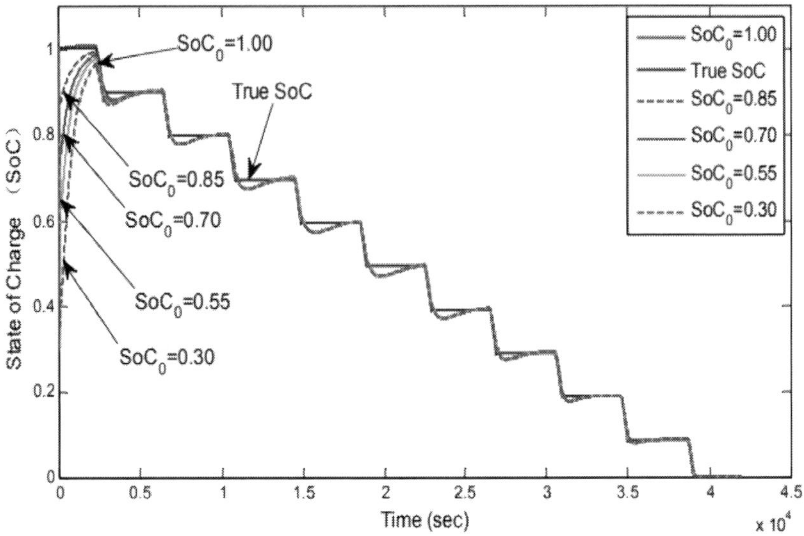

Figure 3. [13] SOC curves for different initial SOC values.

Time Update ('Predict')	Measurement Update ('Correct')
(1) Project the state ahead: $\hat{X}_k^- = A\hat{X}_{k-1} + B_{uk}$ (2) Project the error covariance ahead: $P_K^- = AP_{k-1}A^T + Q$	(1) Compute the Kalman Gain: $K_k = P_K^- H^T(HP_K^- H^T + R)^{-1}$ (2) Update estimate with measurement y_k : $\hat{X}_k = \hat{X}_k^- + K_k(y_k - H\hat{X}_k^-)$ (3) Update the error covariance: $P_k = (I - k_k H)P_K^-$

Figure 4. [17] Nonlinear State Space Model of EKF.

Figure 4 [14] is a group of mathematical equations utilized for the estimation of SOC under nonlinear circumstances.

As mentioned in Lee et al. 2007 [15] and Zhu et al. 2012 [16] - The outcomes of the system using Extended Kalman Filter provides the accurate measurement of %SOC and Ah (Ampere Hour) counting technique in terms of efficiency and dynamic flexibility of a battery.

The calculation steps of his proposed model has been listed in below equations - (1), (2) & (3) - [17].

$$Xk + 1 = f(Xk, uk) + wk \tag{1}$$

$$Yk + 1 = g(Xk, uk) + vk \tag{2}$$

$$\hat{x}_0^+ = E[x_0], P_0^+ = E[(x - x_0^+)(x - x_0^+)^T \tag{3}$$

The eqn. (1) represent the state equation of a system and eqn. (2) represent the output equation of a system. Function $f(Xk, uk)$ *and* $g(Xk, uk)$ is a nonlinear evolution and measurement function; while the *wk* & *vk* represents the process and measurement noise within the system. The EKF starts filtering values from its initial state (x_0^+) and error covariance ($\hat{p}_0^+)$, which is mentioned in eqn. (3).

SOC Estimation Using Unscented Kalman Filter

In EKF, only one point is utilized, i.e., mean and approximate for estimation of nonlinear terms, while the Unscented Kalman Filter uses a group of points known as sigma points and estimate it with help of statistics. The greater number of sigma points provides a highly efficient results which yields in accurate SOC through battery [18].

He et al. 2014 [19] used the UKF method for the SOC estimation of lithium ion battery. The UKF filter for SOC estimation was designed using the posterior mean and Taylor series expansion method. The response of this technique was very accurate in real time domain for determining the nonlinear operating performance of a battery. When UKF implemented for battery cell model, the robustness for faster convergence tracking ability of SOC has been found very accurate. The working structure of this method is shown in Figure 5 [19].

F. Liu [20] used UKF for SOC estimation algorithm-based techniques for the identification of a dynamic parameter in EV batteries. In his findings, the LS-UPF (Least Squared Unscented Particle Filter) has been compared with GA-UPF (Genetic Algorithm and Unscented Particle Filter). The amended GA-UPF process suggested in this research is having great accuracy and precision for the estimation of battery SOC in real time domain.

DATA DRIVEN BASED ESTIMATION METHODS FOR SOC

There has been a significant development in the field of Artificial Intelligence (AI), Data Analytics, Data Science (DS), Machine Learning (ML) and Deep learning (DL) which has led to advancements of different real time, optimized, estimation and predictive analytical systems. Since, there has been a sharp increase in the accuracy of the sensing and sophisticated-testing platforms and large volume of data is made available, the data-driven methods are proven to be more robust and useful [21].

Data driven techniques are widely used in modeling and learning in engineering [22]; which in present context involve the classification and optimization of the various parameters of the battery and finding linear as well as non-linear co-relations between the variables and parameters associated to the battery state. As elaborated in [23], the Data-driven techniques are highly efficient, reliable and unlike filter-based adaptive techniques; they are less-affected by the external unexpected noise in the system if the data-model is being correctly chosen for the application.

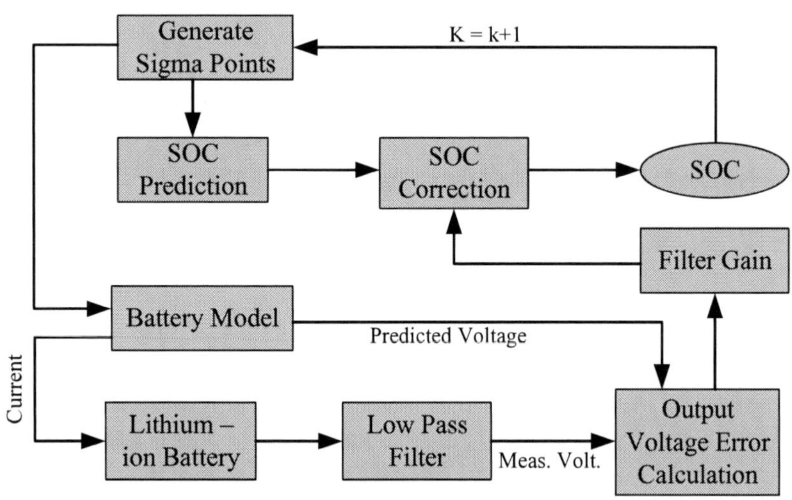

Figure 5. [19] SOC estimation through UKF.

SOC Estimation Using Support Vector Machine (SVM)

Support Vector Machine algorithm is an optimization method to search for an ideal hyperplane in N-dimensional space that can classify as well as predict valid outcomes [24]. In data-driven approach, the features are the input parameters of the battery state which comprises of battery voltage, current value, temperature, cycle number, environmental parameters, etc.

Hyperplanes in SVM model are the decision boundaries that are used to classify the data points into different classes and hence prediction is done accurately by transforming the data to a comparative higher-dimensional space for non-linear data separation by use of kernels. After each iteration, a cost function is computed and from that a new gradient is calculated; and the weights of the consequent parameters of the system is being updated.

SVM model has a few important hyper-parameters which can be tuned as per requirements to achieve higher accuracy such as –

1) Kernels: According to the behavior of the data-points in the dataset; the kernels are be-ing chosen. In battery estimation; Gaussian RBF (radial based function) Kernel and Polynomial Kernel can elevate the accuracy significantly.
2) Regularization ‘C’: The Regularization or the penalty parameter determines how much error is bearable while fitting the model on data.
3) Gamma: The Gamma value defines how far it influences the calculation of plausible line of separation; meaning lesser gamma value influences only near points to the line/hyperplane, while high gamma value influences far points from decision boundary also.

These hyper parameters are hyper tuned by Grid Search techniques [25]. SVM works adroitly only under constant pulse charging/discharging working conditions. Thus, by the various input battery parameters such as temperature, voltage, etc. the SVM model outputs the battery SOC; and therefore, the discharging curve can be predicted.

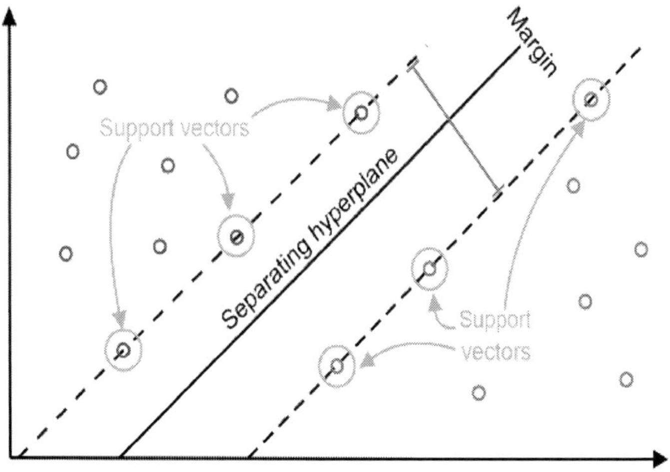

Figure 6. [46] SVM Estimation Technique represented in graphical form for SOC.

Alvarez Anton et al. 2013 [26], has estimated the SOC by using the Support Vector Regression approach for a high capacity lithium iron phosphate (LiFePO4) battery cell. The coefficient of determination as high as 0.97 has been observed during the validation of Support Vector Machine model with new datasets. Also the estimation result was very quick and can be manipulated with any inexpensive digital processor.

Terry Hansen 2016 [27], has also estimated the battery SOC of a large scale lithium-ion-polymer (LiPo) battery using the SVM technique. During the experimentation, the root mean square error has been observed as 5.76%; while the maximum positive and negative error has been observed as +12% and -2% respectively. The overall experimental results culminates that the SOC estimation using SVM requires less computational intensity and provides a faster result in real time domain.

SOC Estimation Using Neural Networks

Neural Network is a mathematical layered model, which is being developed being influenced by working of Human Brain's neural system [28]. It has a series of data processing units called "neurons" or "nodes" stored as "tensors" (a high-dimensional array data structure) which has subsequent weights functions associated with them. Each neuron is a part of different layers, and the values are stored in them as tensors (or vectors). According to the respective mathematical functions and algorithms of the model; after each iteration, the weights of the input neurons are updated by various cost functions, encodings and transformation algorithms; and hence only the optimal neurons contribute to the estimation of the target variable which is the output neurons of the network. Neural Networks are highly robust and stable due to their layered parallel architecture. For the battery state estimation, from [29, 30] it is apparent that Neural Networks prove to be an efficient choice.

For battery state estimation, the idea of the prediction of the SOC value is by inputting the practically calculated battery data. The data is prepared for the training by choosing properly functioning test battery according to the values of upper and lower cutoff voltages, charging methods such as constant current-constant voltage, choosing drive cycles, monitoring computer and data fetcher devices, battery cyclers, temperature measuring devices, tensor flow backend, GPUs, processor units, etc. as used in [31, 32]. Different neural networks such as Fuzzy NNs, Back Propagation NNs, Radial Basis Function NNs, NIO NNs, NARX NNs, RNNs, DNNs, etc. are widely used for this application. Out of the above mentioned, some of the most efficient and latest neural network architectures used for as follows –

Artificial Neural Networks (ANNs)

Artificial Neural Network (ANN) is a high dimensional layered architecture having different layers such as an input layer, hidden layer/s and an output-layer. An input layer is consisting of x-input neurons and each of them is linked to the hidden-layer. In most ANN architectures it is being observed that not more than two hidden layers are used. The nodes of each hidden layer are connected with each other and then is connected to the output layer consisting of y-output layers. In predictive analysis approach, there is only one output neuron in the form of a vector. All the corresponding nodes have respective weights associated with them which are used during the computations that are carried out on training data. It is being trained upon the data collected by various target resources; and after the training process is over, the model can accurately predict the output when it is given completely new data. Based on the architecture, the neural-networks are mainly classified into two types based namely; feed-forward NNs and back-propagation NNs [33].

Ismail M. et al. 2017 [35], used the ANN for estimating the SOC of a lithium ion battery cell. During this experimentation, the performance of ANN has been tested using standard driving cycles. The Root Mean Square Error (RMSE) of 3.85e-09 on the Urban Dynamometer Driving Schedule (UDDS) cycle has been achieved.

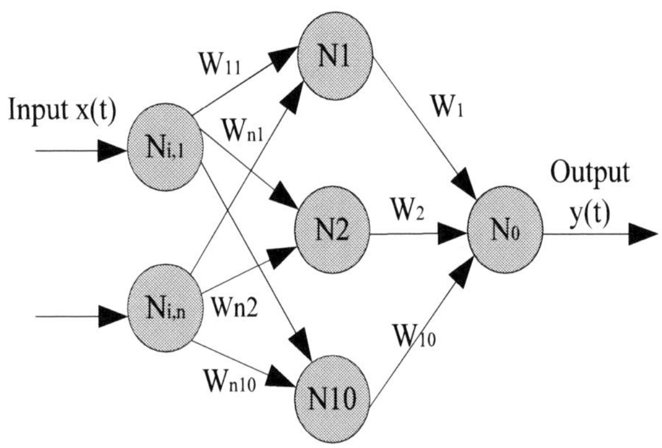

Figure 7. [34] Basic Feedforward ANN.

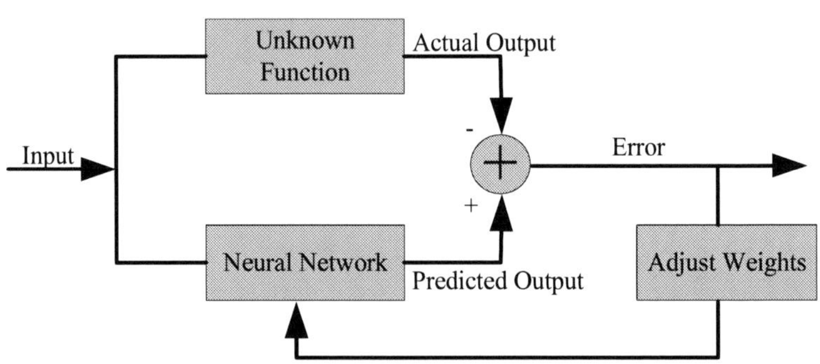

Figure 8. [34] Basic operational block diagram of ANN.

Similarly, S. Grewel (2001) [36] has used the three layer feed forward artificial neural network which has been trained using the back propagation method. During his experimentation, the improvement in accuracy has been achieved by reduction in sum squared error from 0.02 to 0.01 using a 5 neurons at a time.

NIO Feed forward Neural Networks (NIO FFNs)

Non-linear Input/output Feed-forward Neural Network (NIO FFN) [32] is a feed-forward neural network with open loop architecture. It estimates the input values of present as well as previous values of given time-series data. The Input layer gets the data values with delay parameter associated with the series of input values from the data. The input layer has different 'features' for Battery Estimation such as terminal voltage and current. The data passes through hidden layer and returns the SOC at the output layer [37]. Each neuron undergoes several operations such as activation functions, transformation, etc. and then forwarded to the other subsequent layers. The mathematical form of this network can be represented as:

$$y(t) = f(x(t-1), \dots\dots\dots, x(t-d)) \tag{4}$$

where, y(t) is output and x(t) denotes the input parameter values applied under the function 'f.'

NARX Feed Back Neural Networks (NARX BPNs)

Non-linear Autoregressive with Exogenous input (NARX) Back Propagation Network (BPNNs) is a neural network consisting of a closed loop having capabilities to optimize the loss function by learning from the calculated errors and updating the weights after each series of input data by making use of back-propagation algorithm [37, 38]. The weights associated with each neuron is optimized to develop an ideal estimation system based on gradient descent approach. The NARX feedback Neural Network predicts one time series on the basis of the input of present and previous values of input time-series of data along with the previous and present time series learnings (as it back propagates each iteration) and hence it is called external/exogenous time series output. The output equation is given as:

$$y(t) = f\big(x(t-1), \dots, x(t-d)\big), y(t-1), y(t-2) \dots, x(t-d) \tag{5}$$

where, y(t) is output at time t, x(t-i) are input parameters and y(t-i) are previously obtained output values which are backpropagated applied under function 'f.'

Since NARX neural networks have input based on time series data, they are often classified as a subclass of Recurrent Neural Networks [31]. It uses feedbacks from the estimated %SOC value from the out layer which increases the learning capacity, convergence speed and accuracy.

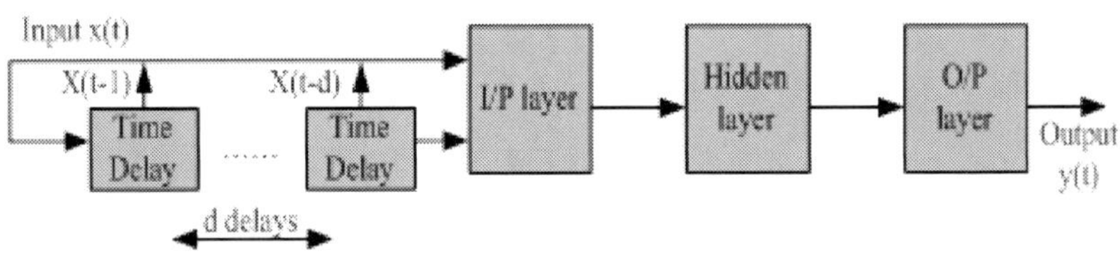

(a)

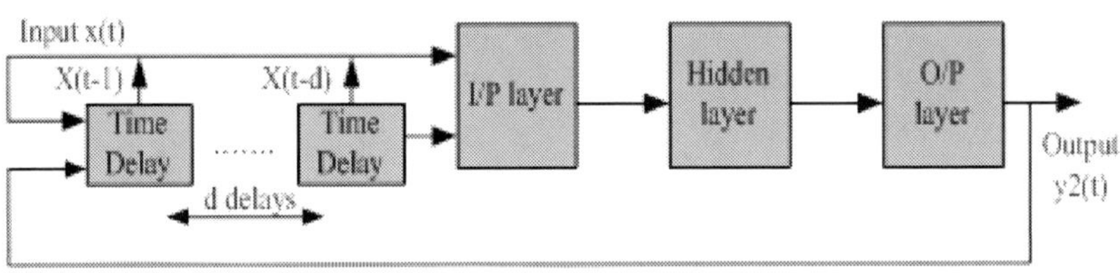

(b)

Figure 9. (a) [34] The architecture and working of the NARX and NIO NNs. (b) [34] The architecture and working of the NARX and NIO NNs.

Fine Tuning of the Artificial Neural Networks (ANNs) and Comparative Analysis

The process that is used to increase the efficiency and speed up the training time for the neural networks by using various techniques is called Fine Tuning. It comprises of defining update rate of the gradients, choosing proper activation function, adding penalties and regularization methods [39], early stopping techniques [40], finding optimal dimensions of hidden layer, etc. While for the SOC estimation, the data input is of time-series type and hence; the delay parameter plays a very important role. By choosing the most optimum value for the delay ‘d’ by observation of the accuracies, one can choose the correct delay for the input series data. A. A Hussain [34] conducted practical experiments for prediction of SOC on NIO and NARX neural networks and from those experiments, various trends in the SOC estimation of NIO and NARX were observed. In NIO, with the increase in delay parameter ‘d’, the fluctuations in the predicted SOC value from the original SOC value decreases.

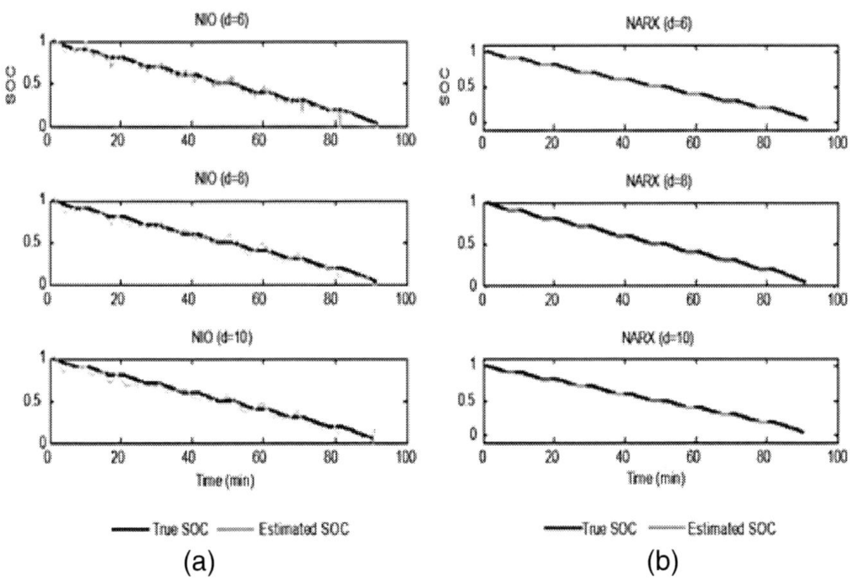

Figure 10. The graphical representation of experimental results produced in [34] by using NIO and NARX NNs.

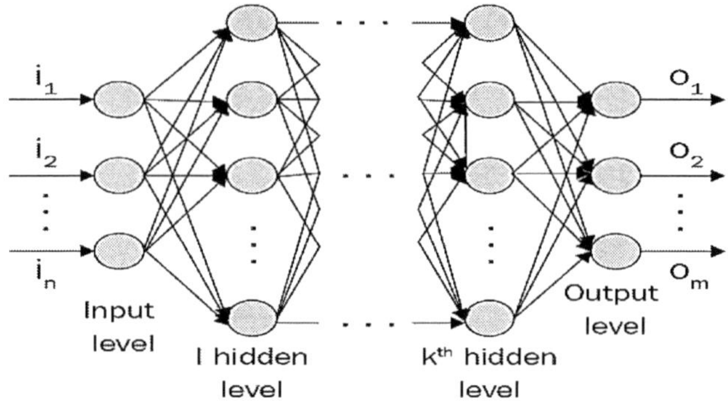

Figure 11. [57] DNN Architecture.

While, for the NARX, almost negligible noise is encountered with the changes in the delay and remained unaffected. The NARX outperformed NIOs due the presence of feedback loop system. On fine tuning of parameters such as delay, etc., it was practically found that the NIO neural networks can decrease the Root Mean Squared Error (RMSE) to about 0.0379 by rigorous tuning. On the other hand, NARX model has most optimal RMSE value around 0.0074, which is highly reliable. Also improved version of BP neural network is used for better results [41].

Deep Feedforward Neural Networks (DNNs)

It refers to a large network architecture of the neurons having multiple hidden layers between the input and the output layers. Deep Feedforward Neural Networks have been used for the highly precise approximation of the SOC. The discussed model as in [32] has the four input parameters as 4 input layers: namely Average current, Voltage, temperature and average voltage of the battery.

The DNN architecture, as shown in Figure 11 having 'n' input neurons, 'k' hidden layers (a1, a2, a3 … ak number of neurons in each corresponding hidden layer) and an output layer of 'm' neurons.

Modern research has significantly reflected the pros of using deep learning with the existing estimation techniques. DNN has multiple hidden layers whose terminal layers connected to input and output respectively.

Chemali et al. 2018 [32] carried out the training of the battery data on DNN architecture model. In that research, apart from the DNN model, various other SOC estimation models such as Adaptive-improved EKF, AUEKF with extreme Machine Learning model, AEKF with ANN model, Fuzzy Neural Network (FNN) with Genetic Algorithm, Radial Bias Function NN (RBFNN) were also studied. By the application of drive cycle loads on the lithium-ion battery at different ambient temperatures, the training data is generated. The dynamic drive cycles were in the range of ambient temperature having the current ratings in ±18 A. To produce variable and dynamic dataset for the battery estimation, the temperature parameter is being varied from -20°C to 25°C. The mapping of the input signals from the battery is done very efficiently by the DNNs and provides high accuracy with highly optimized Mean Absolute Error (MAE) as 1.10% for data generated at 25°C and 2.17% over -20°C dataset.

Recurrent Neural Networks (RNNs)

With the advent of development in the field of deep-learning; one of the most effectual method for dealing with the time-series data is Recurrent Neural Network (RNN) [42, 43]. Since the artificial neural networks depend only on the variables of a current state, the effect of the variables in the past state is neglected on the system. Thus, the trained model cannot accurately predict the future states. RNNs can use its memory state to learn the parameters and their consequent time-dependencies obtained from the sequentially passed data. According to architecture, RNN Cells are classified into three major types: -

1) Vanilla/Simple RNN Cell
2) GRU (Gated-Recurrent Unit) Cell
3) LSTM (Long Short-Term Memory) Cell [44].

Any RNN network can be seen as a series of recurrent cells as described in [45]; which has input gates for passing inputs to hidden layers in which different operations like sigmoid, tanh, transformations are carried out. After this, the outputs are forwarded to next cell via the output gates. Simple/Vanilla RNN Cell structure is a light-weight model consisting of only one activation function, i.e., tanh for estimation of the output according to the provided input. While in GRU cells, two gates are used namely reset and update. LSTM is more complex architecture, where three gates are used namely Input, Forget and Output gates are introduced. Since, GRUs are less complex; they require relatively less training time than Simple RNNs and are faster; contrary the LSTMs require high computational capabilities, but they are highly accurate for a long sequential data.

By the use of RNNs, the mapping of battery parameters like temperature, voltage, current, etc. is done directly to the SOC estimation and avoids the demerits of filter-based methods like sensitivity towards noise, accurate battery, computationally expensive in real time, etc. RNNs perform self-learning by back propagation in each cell by using stochastic gradient algorithms; while in filter-based methods there is a challenge of parametrizing the model correctly. Moreover, RNN can learn to estimate SOC on different network as well as varying environmental conditions. The training along with the valuable outputs of the GRU Cells was conducted by Li et al. 2019 [46]. Also, in the field of RNN for SOC estimation, the behavior of LSTM cells has been studied by Chemali et al. 2018 [47].

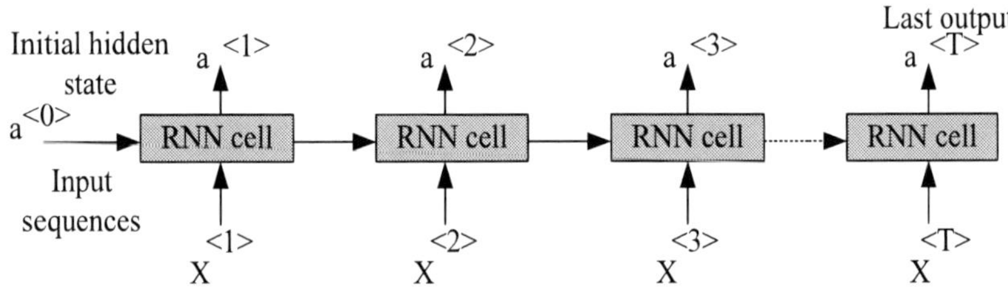

Figure 12. [51] SOC Architecture of a Recurrent Neural Network.

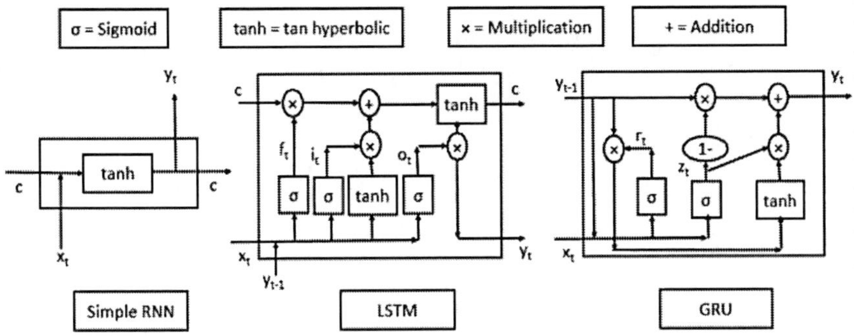

Figure 13. [52] Architecture of Simple RNN, LSTM and GRU cells.

From the findings, it is concluded that both have their respective pros and cons; and hence must be chosen according to the desired estimation accuracy and response-time. The RNN architecture & cell-architectures has been mentioned in Figure 12 and 13 respectively. Moreover, the graphical outcomes of the experimentations are also shown in the Figure 14 and 15 respectively.

In the Figure 14, the %SOC values and %SOC error are plotted under high-pulse discharge conditions at different temperatures such as: 0°C, 25°C & 45°C respectively. Similarly, in Figure 15, the graphs of the %SOC values and %SOC errors are represented at different ambient temperatures such as: 25°C, 10°C & 0°C respectively.

Reinforcement Learning (RL) Approach

The Data-driven techniques are more powerful when there is availability of large amount of accurately measured data for a specific application. But sometimes due to less amount of past data, and most of 'target' data is in real time; a dynamic data processing system is considered beneficial. The Reinforcement Learning is a branch of Machine-Learning, where the predictions are made in real time and after some iterations; the accurate estimations are obtained. The SOC estimation by Reinforcement (or Q-Learning) approach is a Hybrid-estimation method, combining both filter-based and data-driven estimation methods.

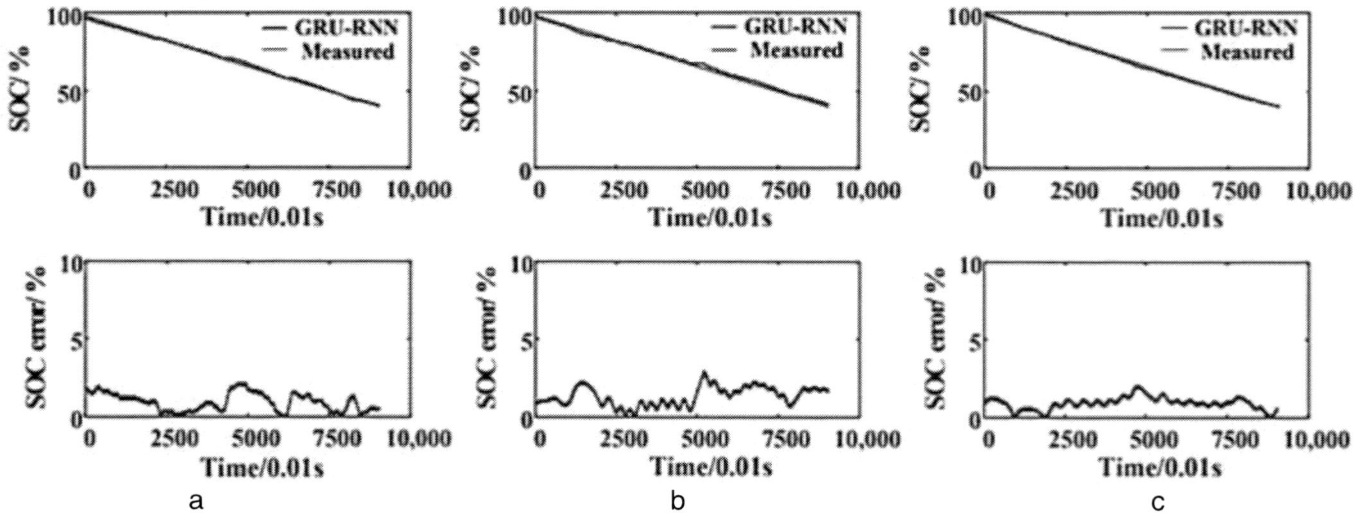

Figure 14. %SOC curves and %SOC error graphs of GRU RNNs [46].

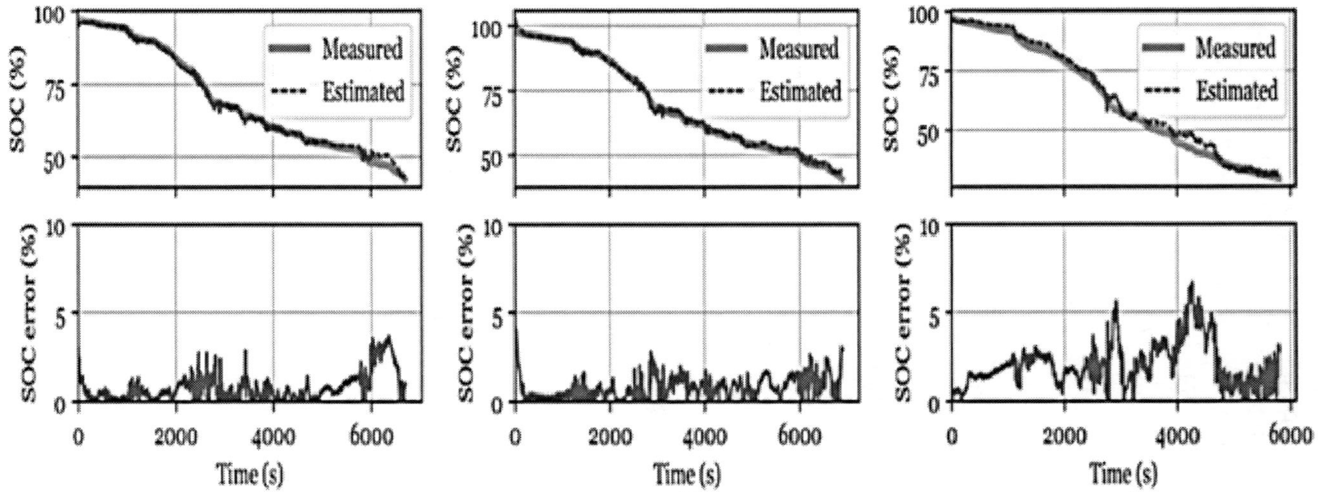

a b c

Figure 15. %SOC curves and %SOC error graphs of LSTM RNNs [47].

The working of RL comprises of some target 'agent' in some environment 'e,' in a state 's' performs an action 'Q' (Q-value). According to the defined policy (p) of the chosen model, a reward (R) is given immediately after it performs an action generated by Value Function. After a certain time of its operation, it generates a value (V) which determines long-term return by the model in performing a particular task. Each iteration is called as 'episode' in Reinforcement learning.

Kim et al. 2018 [48] drew valuable conclusions by performing the training process of the RL-based SOC estimator. In the MDP approach, the iterative EKF is used with 1RC equivalent circuit model and RL is used for the parameter optimization. By making use of EKF, considerable accuracy is achieved and at the same instance the proposed RL model operates through parameter adjusting rule. The parameter adjusting rule changes the EKF parameters reference to the calculated voltage, current, etc. [49, 50]. The battery parameters such as usage pattern of battery, error between model and real battery [51,52], aging cycles and initial error value of estimation apart from the voltage and current values has also been incorporated within his research [53,54]. For SOC estimation, Kalman Filter family (KF, EKF, etc.) can be taken as actions and states of a system as a model-based technique and data-driven technique as RL models like MDP [55], Q-learning can be used to adjust rules, pro-vide a rewards and henceforth, estimating the SOC of a system [56].

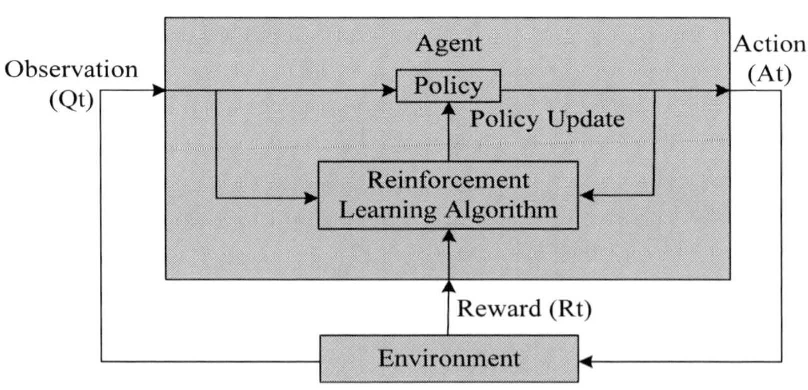

Figure 16. [58] Reinforcement Learning Flow-diagram.

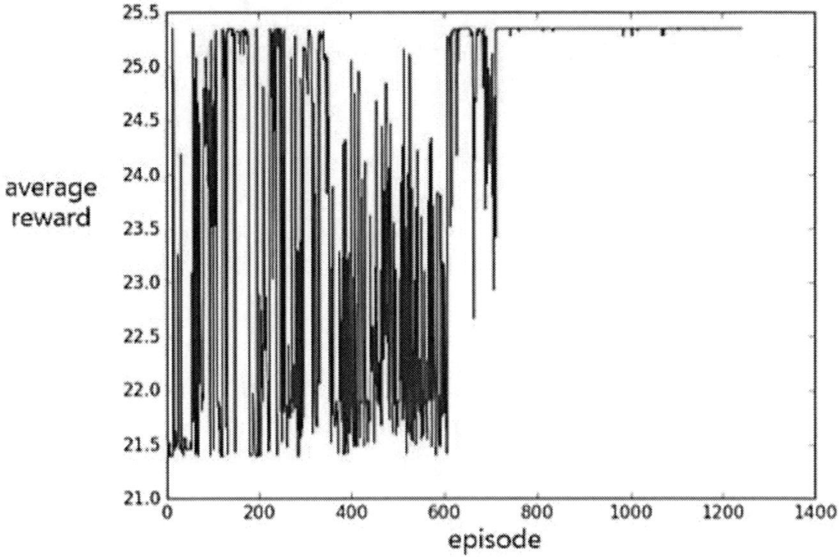

Figure 17. [52] Average reward value plotted against each episode (iteration) according to experimental results.

The basic flow-diagram of Reinforcement Learning [52] process is as shown in the Figure 16. From the experiments conducted in [57], the conclusive graph of the average reward versus the episode (iteration) has also been mentioned in Figure 17.

Future Works in SOC

After analyzing many published papers, articles, patents and sundry data sheets of different battery packs, it has been visually examined that the future of SOC estimation technology requires various types of research to enhance its precision level. Some of the future work for enhancing the precision and accuracy level of SOC is mentioned as under –

1) The efficiency and stability of the model must be improved without increasing the intricacy and computational intensity [58, 59].

2) In filter predicated approach, the accurate optical discernment of process and quantification noise is mandatory.
3) The Particle Filter works very well for the estimation of nonlinear datasets, but the quandary is its highly mathematically computational equations which increases its intricacy level. Due to this, the EKF and SPKF has been utilized, which has a good estimation in authentic time domain, but not better than Particle Filter. Hence, more research is required to minimize its intricacy, so that it could be possible to implement for %SOC estimation in real time domain.
4) In order to amend the precision of the EKF, threshold signal detector technique along with felicitous monitoring of Gaussian noise can be implemented in future.
5) The focus should be laid more on designing of better neural network architectures with lesser complexity and higher adaptive learning capabilities. According to the desideratum, the conception should be towards development of more perspicacious algorithms to solve the involute quandaries.
6) More research in the hybrid techniques should be carried out for the SOC estimation. Since the Hybrid estimation methods includes the use of different model-based, filter-based or data-driven techniques in the designof the estimator. The model-based or filter-based techniques are less complex, while the data-driven methods are highly complex but more accurate compared to others, so integrating them can result into a more stable and robust system.
7) Discovering of the other possible significant battery parameters for the SOC estimation which are derived from the aging cycles, degradation cycles, discharging curves, etc. is an imperative research direction. By the use of modern machine learning algorithms and statistical analytics, the new parameters can be derived and can predict SOC in a much efficient way.
8) In the neural network approach, the training time can be truncated significantly by fine tuning of neural networks. Additionally, research to find optimal number of neurons in consequent layer, appropriate activation functions, weight-update policies for particular SOC estimation can be tuned and

tested to obtain more precise and more expeditious replications.

CONCLUSION

This paper produces a valuable cumulation of essential and advanced estimation methods for the SOC of battery. Different Filter based and Data-driven estimation methods such as - KF, EKF, UKF, ANN, RNN, NARX, NIO, SVM, DNN, etc. along with their mathematical computational equations, architecture, parameters, functions and operations for the real time implementation in battery model are being explicated in detail. Also, the experimental conclusions including error values, graphs, models, etc. on different research papers of corresponding topics have been explicated. Due to nonlinear variations in battery cell model, EKF and UKF is used in place of linear (simple) KF. The Analytics Linearization and Statistical/Empirical Linearization model at each point in time increases the accuracy of EKF and UKF in real time domain [55]. Despite of unavailability of accurate battery and noise models in certain scenarios, the data-driven approach proves to be an efficient solution due to their highly stable network architecture. The hybrid estimation approach of Reinforcement learning is also explained which amalgamates the filter and data driven techniques into meticulous estimations. Also, as an important note for the further development in the present topic, the future research areas and improvements for advancement in the SOC estimation field are being deliberately discussed.

CONFLICT OF INTEREST

This research did not receive any specific grant from funding agencies in the public, commercial, or not for profit sectors.

ACKNOWLEDGMENTS

The author would relish to thank the Microsoft teams, for providing the license free Microsoft Visio 2007 SP1 version for engendering the sundry flowcharts and diagrams.

REFERENCES

[1] Hook, Mikael, and Xu, Tang. 2013. "Depletion of fossil fuels and anthropogenic climate change - A review." *Energy Policy* 52: 797-809.

[2] Oliveira, Beatriz Fátima Alves de, Eliane, Ignotti, and Sandra S. Hacon. 2011. "A systematic review of the physical and chemical characteristics of pollutants from biomass burning and combustion of fossil fuels and health effects in Brazil." *Cadernos de Saude Publica* 27(9): 1678-1698.

[3] Yeau, C. W. 2013. "The State of Charge Estimating Methods for Battery: A Review." *ISRN Applied Mathematics* 2013: 1-7.

[4] Bart, B., and Kees, Z. 2016. "Model based estimation at Statistics." *Netherland, European Conference on Quality in Official Statistics* Q2016: 1-10.

[5] Jinhao, M., et al. 2019. "A Simplified Model based State-of-Charge Estimation Approach for Lithium-ion Battery with Dynamic Linear Model." *IEEE Transactions on Industrial Electronics* 66(10):1-8.

[6] Prashant, Shrivastava, Tey Kok Soon, Mohd Yamani Idna Bin Idris, and Saad, Mekhilef. 2019. "Overview of model-based online state-of-charge estimation using Kalman filter family for lithium-ion batteries." *Renewable and Sustainable Energy Reviews* 113.

[7] Zhang, J., and Lee, J. 2011. "A review on prognostics and health monitoring of Li-ion battery." *Journal of Power Sources* 196(15): 6007-6014.

[8] He, W., et al. (2013). "State of charge estimation for electric vehicle batteries using un-scented kalman filter." *Microelectronics Reliability* 53(6): 840-847.

[9] Mastali, M., et al. 2013. "Battery state of the charge estimation using Kalman filtering." *Journal of Power Sources* 239: 294-307.

[10] Gregory, Plett. 2004. "Extended Kalman filtering for battery management systems of LiPB-based HEV battery packs: Part 1. Background." *Journal of Power Sources* 134(2): 252-261.

[11] Lee, J., Nam, O., and Cho, B. H. (2007). "Li-ion battery SOC estimation method based on the reduced order extended Kalman filtering." *Journal of Power Sources* 174(1): 9-15.

[12] Kim, Jonghoon, et al. 2012. "OCV hysteresis effect-based SOC estimation in extended Kalman filter algorithm for a LiFePO4/C cell." In: *2012 IEEE International Electric Vehicle Conference*: 1-5.

[13] Yuan, S., Wu, H., and Yin, Y. 2013. "State of Charge Estimation Using the Extended Kalman Filter for Battery Management Systems Based on the NARX Battery Model." *Energies* 6: 444-470.

[14] Grewal, M. S., and Andrews, A. P. 2001. *Kalman Filtering: Theory and Practice using MATLAB. 2nd edn*. John Wiley & Sons.

[15] Lee, S. J. et al. 2007. "The state and parameter estimation of an Li-Ion battery using a new OCV-SOC concept." In: *Proceedings of the 2007 Power Electronics Specialists Conference*: 2799-2803.

[16] Zhu, Z. et al. 2012. "Online state of charge EKF estimation for LiFePO4 battery management systems." In: *Proceedings of the International Symposium on Intelligent Signal Processing and Communications Systems*: pp. 609-614.

[17] Sepasi, S., Ghorbani, R., and Liaw, B. Y. 2014. "Improved extended kalman filter for state of charge estimation of battery pack." *Journal of Power Sources* 255: 368-376.

[18] TOWARDSDATASCIENCE ekf. 2021. *Data science.* Accessed March 20. https://towardsdatascience.com/the-unscented-kalman-filter-anything-ekf-can-do-i-can-do-it-better-ce7c773cf88d.

[19] He, H., et al. 2014. "Lithium ion battery SOC estimation with UKF and RTOS µ COS - II Platform." *Energy Procedia* 61: 468-471.

[20] Liu, F., Ma, J., and W, Su. 2019. "Unscented Particle Filter for SOC Estimation Algorithm Based on a Dynamic Parameter Identification." *Hindawi*: 1-14.

[21] Li, Y. et al. 2016. "Dynamic data-driven and model-based recursive analysis for estimation of battery state-of-charge." *Applied Energy*. 184: 266-275.

[22] Montans, F. J. et al. 2019. "Data driven modeling and learning in science and engineering." *Comptes Rendus Mecanique* 347(11): 845-855.

[23] Venugopal, P. and Vigneswaran, T. 2019. "State-of-Charge Estimation Methods for Li-ion Batteries in Electric Vehicles." *International Journal of Innovative Technology and Exploring Engineering (IJITEE)* 8(7): 37-47.

[24] Basak, D. et al. 2007. "Support Vector Regression." *Neural Information Processing - Letters and Reviews* 11(10): 203-224.

[25] Syarif, I., et al. 2016. "SVM Parameter Optimization using Grid Search and Genetic Algorithm to Improve Classification Performance." *TELKOMNIKA (Telecommunication Computing Electronics and Control)* 14(4): 1502-1519.

[26] Anton, J. C. Alvarez., et al. 2013. "Battery state-of-charge estimator using the SVM technique." *Applied Mathematical Modelling* 37 (9): 6244-6253.

[27] Hansen, Terry, and Chia-Jiu, Wang. 2005. "Support vector based battery state of charge estimator." *Journal of Power Sources* 141: 351-358.

[28] Pathmind, N. N. 2020. *Neural Network*. Accessed July 19. https://pathmind.com/wiki/neural-network/neuralnetwork.

[29] Bermejo, D. J., et al. 2018. "Using Dynamic Neural Network for Battery State of Charge Estimation in Electric Vehicles." *Procedia Computer Science* 130: 533-540.

[30] Jayasinghe, J. A. K. S., et al. 2014. Neural Network Based State of Charge (SOC) Estimation of Electric Vehicle Batteries. *International Journal of Scientific and Research Publications* 4(12) 2250-3153.

[31] Lipu, M. S. H., et al. 2017. "State of Charge Estimation for Lithium-ion Battery Using Recurrent NARX Neural Network Model Based Lighting Search Algorithm." *IEEE Access* 6: 28150-28162.

[32] Chemali, E., et al. 2018. "State-of-charge estimation of Li-ion batteries using deep neural networks: A machine learning approach." *Journal of Power Sources* 400: 242-255.

[33] Enache, B. A., and Diaconescy, E. 2014. "Estimating a battery State of Charge using neural networks." *International Symposium on Fundamentals of Electrical Engineering* 1-06.

[34] Hussain, A. A. 2014. "Kalman Filters versus Neural Networks in Battery State-of-Charge Estimation: A Comparative Study". *International Journal of Modern Nonlinear Theory and Application*. 3 (5): 199-209.

[35] Ismail, M., et al. 2017. "Battery state of charge estimation using an Artificial Neural Network." In: *2017 IEEE Transportation Electrification Conference and Expo (ITEC)*: 342-349.

[36] Zeynali, Saeed, Naghi, Rostami, Ali, Ahmadian, and Ali, Elkamel. 2020. "Two-stage stochastic home energy management strategy considering electric vehicle and battery energy storage system: An ANN-based scenario generation methodology." *Sustainable Energy Technologies and Assessments* 39: 1007-1022.

[37] Yurkiv, Vitaliy, et al. 2018. "The influence of stress field on Li electrodeposition in Li-metal battery." *MRS Communications* 8 (3).

[38] Yannlecun Back Prop. 2021. *Back Propagation*. Last accessed September 20. http://yann.lecun.com/exdb/publis/pdf/lecun-98b.pdf.

[39] Cai, Chenghui, et al. 2002. "State-of-charge (SOC) estimation of high power Ni-MH re-chargeable battery with artificial neural network." In: *Proceedings of the 9th International Conference on Neural Information Processing 2002ICONIP'02* vol. 2: 824-828.

[40] Cai, C. H., et al. 2003. "Battery state-of-charge (SOC) estimation using adaptive neuro-fuzzy inference system (ANFIS)." In: *The 12th IEEE International Conference on Fuzzy Systems*, 2003. FUZZ'03 (2):1068-1073.

[41] Guo, Y., Zhao, Z., and L, Huang. 2017. "Soc estimation of lithium ion battery based on improved BP neural network." *Energy Procedia* 105: 4153-4158.

[42] Abbas, Ghulam, Menaa, Nawaz, and Farrukh, Kamran. 2019. "Performance comparison of NARX & RNN-LSTM neural networks for LiFePO4 battery state of charge estimation." In: *2019 16th International Bhurban Conference on Applied Sciences and Technology (IBCAST):* 463-468.

[43] Jiao, Meng, Dongqing, Wang, and Jianlong, Qiu. "A GRU-RNN based momentum optimized algorithm for SOC estimation." *Journal of Power Sources* 459: 228-251.

[44] Song, Xiangbao, et al. 2019. "Combined CNN-LSTM network for state-of-charge estimation of lithium-ion batteries." *IEEE Access* 7.

[45] Karpathy, A., Johnson, J., and Fei, L. F. 2015. "Visualizing and understanding recurrent networks." arXiv.

[46] Li, B. S., Xiao, F., and Fan, Y. 2019. "An Approach to State of Charge Estimation of Lithium-Ion Batteries Based on Recurrent Neural Networks with Gated Recurrent Unit." *MDPI* 12(9).

[47] Chemali, E., et al. 2018. Long Short-Term Memory-Networks for Accurate State of Charge Estimation of Li-ion Batteries. In: *IEEE Transactions on Industrial Electronics* 65(8):1-10.

[48] Kim, M., et al. 2018. "State of charge estimation for lithium ion battery based on reinforcement learning." *IFAC Paper Online* 51(28): 404-408.

[49] Reddy, Namireddy Praveen, et al. 2019. "An intelligent power and energy management system for fuel cell/battery hybrid electric vehicle using reinforcement learning." In: *2019 IEEE Transportation Electrification Conference and Expo (ITEC)* 20: 1-6.

[50] Packtpub SVM. 2020. *Support Vector Machine*. Accessed November 2020. https://subscription.packtpub.com/book/big_data_and_business_intelligence/9781789345070/3/ch03lvl1sec30/svm-for-churn-prediction.

[51] S.Kannadhasan and R.Nagarajan, Development of an H-Shaped Antenna with FR4 for 1-10GHz Wireless Communications, *Textile Research Journal*, DOI: 10.1177/00405175211003167 journals.sagepub.com/home/trj, March 21, 2021.

[52] Packthub DNN. 2020. *DNN for SoC*. Accessed November 2020. https://subscription.packtpub.com/book/big_data_and_business_intelligence/978179346640/2/ch02lvl1sec21/building-a-deep-neural-network-to-improve-network-accuracy.

[53] Dlology SOC. 2020. *Sequences.* Accessed November 2020. https://www.dlology.com/blog/how-to-use-return_state-or-return_sequences-in-keras/.

[54] S. Kannadhasan, G. Karthikeyan and V. Sethupathi, *A Graph Theory Based Energy Efficient Clustering Techniques in Wireless Sensor Networks.* Information and Communication Technologies Organized by Noorul Islam University (ICT 2013) Nagercoil on 11-12 April 2013, Published for Conference Proceedings by IEEE Explore Digital Library 978-1-4673-5758-6/13 @2013 IEEE.

[55] S. Kannadhasan and R. Suresh, EMD Algorithm for Robust Image Watermarking. *Recent Advances in Mechanical Engineering and Interdisciplinary Developments Organized by Ponjesly College of Engineering (ICRAMID 2014)* Nagercoil on 7-8 March 2014, Published for Advanced Materials Research Vols.984-985 (2014) PP 1255-1260, ISSN No:1022-6680.

[56] Khan, Z., et al. 2019. "Development and evaluation of recurrent neural network-based models for hourly traffic volume and annual average daily traffic prediction." *Transp. Res. Rec.* 2673(7).

[57] S. Kannadhasan and R. Nagarajan, Performance Improvement of H-Shaped Antenna With Zener Diode for Textile Applications, *The Journal of the Textile Institute*, Taylor & Francis Group, DOI: 10.1080/00405000.2021.194452.

[58] MathWorks RL. 2020. *Reinforcement Learning for SoC.* Accessed November 2020. https://www.mathworks.com/help/ reinforcement -learning/ug/create-agents-for-reinforcementlearning.html.

[59] Rhudy, Matthew, Yu, Gu, and Marcello, R. Napolitano. "An analytical approach for comparing linearization methods in EKF and UKF." *International Journal of Advanced Robotic Systems* 10(4): 208-21.

In: Research Challenges in Science ... ISBN: 978-1-68507-008-3
Editors: S. Kannadhasan et al.

Chapter 11

A Superlative Set Theory for the Prediction of Diabetic Risk

M. Durgadevi*
Department of Computer Science and Engineering,
SRM Institute of Science and Technology,
Faculty of Engineering and Technology,
Vadapalani Campus, Chennai, TamilNadu, India

Abstract

In an actual risk-assessment process enormous classification and prediction algorithms have been developed in the field of data mining, among which only a few have employed mathematical predictions. This chapter introduces an innovative mathematical prediction model to determine the risk level of diabetes by considering different factors of diabetic risk such as Glycosylated Haemoglobin (HbA1c), Plasma Glucose-Fasting (PG-F), Plasma Glucose-Post Prandial (PG-PP) and age with reference to fuzzy set t heory and soft set theory. The real time diabetic dataset from a multispecialty diabetic clinic was used for risk analysis. The predictive ability of the superlative set theory was compared with the Grey prediction, Rough Set Theory and Fuzzy Set theory. The

* Corresponding Author's E-mail: durgadem@srmist.edu.in.

Superlative Set theory has reduced the unnecessary, irrelevant and redundant features in the datasets and has produced superior forecasting results in terms of accurate diabetic risk prediction.

Keywords: fuzzy set theory, soft set theory, superlative set theory, diabetic risk, grey prediction, rough set theory

INTRODUCTION

A variety of algorithms have been used for the disease prediction task as the accuracy of the algorithm is heavily based on the type of data [1]. The most commonly used classification algorithms with regard to Type II diabetes prediction are the decision trees, ANN and SVM. It is worth mentioning that a vast majority of existing approaches enhanced the classification accuracy above 80% in the prediction of Type II diabetes. On the other hand, several healthcare organizations changed their legacy systems to include electronic health records (EHRs) [2]. This significant transformation in the healthcare sector has benefited the administration personnel, physicians, patients and nurses with timely access to medical records when needed. The process of data mining is capable of finding valuable and new data from the huge volume of data [3]. In healthcare, the data mining techniques are widely employed for the prediction of different diseases and for supporting doctors in creating the medical decision during diagnosis.

LITERATURE SURVEY

In the fast-growing world, the occurrence of the diseases increases rapidly and data mining applications are found to be very useful in assisting it. As numerous disease prediction models exist in the literature, it is difficult to explore and review the vast knowledge in the area of healthcare [4]. A study reported that diabetes is considered as a silent killer in India and almost seven million people in India suffer from diabetes [5]. The disease prediction models are extremely useful in various scenarios, via disease diagnosis, personalized medicine, drug discovery and so on. An important application of ML techniques in the medical domain is the process of identifying the disorders which are generally difficult to diagnose [6]. It includes several diseases namely, diabetes, cancer and hepatitis in the earlier stages to other genetic disorders [7]. Another essential application of ML present is the process of discovering drugs at the beginning level of the disease [8]. It includes Research and Development (R&D) technologies, namely, future generation sequencing and precision medicine that helps to find substitutes for the treatment of multi-factorial diseases [9]. Presently, the ML approaches follow unsupervised learning which identifies that the patterns in the data exist without predictions [10]. The behavioral modification plays a major role in the preventive medicine [11]. Many researches have been carried out in the area of cancer prevention and identification, patient treatment, and so on by the development of ML [12]. A data analytics firm, Somatix announced an ML-based application to identify the gestures of the people in day- to-day life to recognize the human unconscious behavior and provides effective solutions [13].

The major intention of ML in medical domain is to reduce time, cost and work [14]. Presently, crowd sourcing has become more important in the healthcare domain which enables academicians and industrialists to access massive quality of data provided by the people based on their own consent. The recent advances made in Internet of Things (IoT) [15] have directed the health care sector to explore various ways of utilizing the data and to handle the tough-to-diagnose cases. It also helps to improve the overall diagnosis and medication.

SUPERLATIVE SOFT SETS RULE MINING

The area of medical diagnosis involves many risk factors which influence the decision-making process by comparing the results with the doctor's opinion. Nowadays, the Artificial Intelligent techniques have been widely used to detect the ambiguity and inarticulacy. The real data set holds the values of Glycosylated Haemoglobin (HbA1c), Plasma Glucose – Fasting (PG-F), Plasma Glucose-Post Prandial and age data of patients which are given as input into the Superlative prediction system and the results are estimated in terms of diabetic risk percentage. The Integrated Development Environment (IDE) Microsoft visual studio and C sharp multi paradigm programming language is used to design the superlative expert system. The procedure involved in the design of the proposed soft expert system is depicted in Figure 1.

Definition I

Consider a set of parameters 'S' belongs to a universal set 'U', then a pair (L, S) is treated as a soft set over U, where L is mapped and represented as $L: SP(U)$.

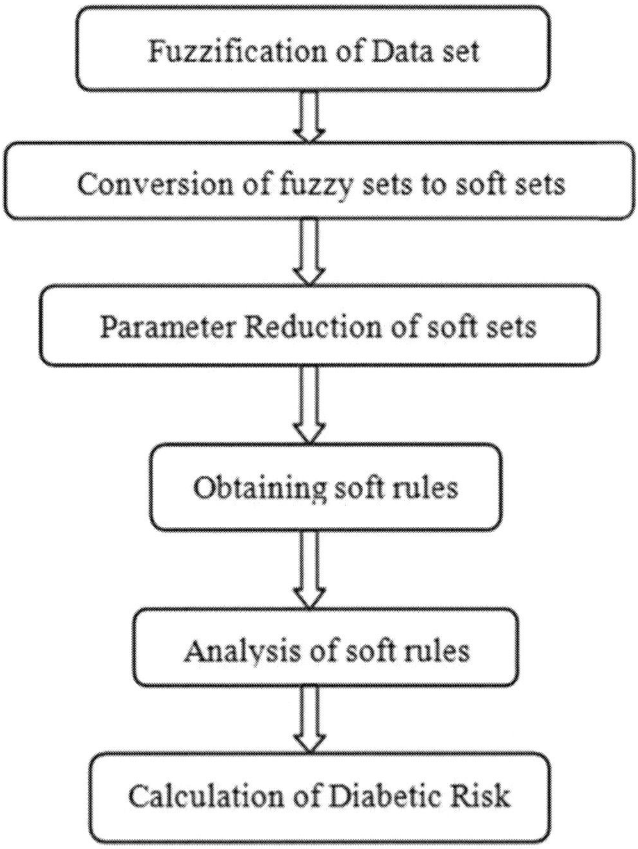

Figure 1. Overall process of Superlative Set based disease prediction.

Illustration

A set of patients who suffer from severe illness need to be admitted in the hospitals located in a nearby place. The facility of the patient's rooms is computed by the use of soft set. The soft set (L, S) displays the room count present in the hospital in the nearby place.

Consider $U = \{P_1, P_2, .., P_6\}$ are the hospital rooms by considering $S = \{S_1, S_2, .., S_5\}$ is the parameter set.

$$L(S_1) = \{P_2, \quad P_4\}$$

$$L(S_2) = \{P_1, \quad P_3, \quad P_4\}$$

$$L(S_3) = \emptyset$$

$$L(S_4) = \{P_1, \quad P_3, \quad P_5\}$$

$$L(S_5) = \{P_1, \quad P_6\}$$

The soft set (L, S) is indicated as follows:

$$(L, \quad S) = \{S_1 = \{P_2, \quad P_4\}, \quad S_2 = \{P_1, \quad P_3, \quad P_4\}, \quad S_3 = \emptyset, \quad S_4 = \{P_1, \quad P_3, \quad P_5\}, S_5 = \{P_1, \quad P_6\}\}$$

The superlative Sets are represented in Table 1.

Table 1. Representation of superlative sets

U	S_1	S_2	S_3	S_4	S_5
P_1	0	1	0	1	1
P_2	1	0	0	0	0
P_3	0	1	0	1	0
P_4	1	1	0	0	0
P_5	0	0	0	1	0
P_6	0	0	0	0	1

Definition II

Consider (E,X) and (H,Y) as the two soft sets in a universal set U. The intersection of (E,X) and (H,Y) is given as $(E,X) \cap (H,Y)$ and represented as a soft set $(I,\ Z)$, where $Z = X \cap Y$.

$$H(E) = \begin{cases} E(E) IFE \in X - Y \\ H(E)\ IFE \in Y - X \\ E(E) \cap G(E) IFE \in X \cap Y \end{cases} \quad (1)$$

Generally, the membership functions (MFs) are employed to encode and decode the variables in a fuzzy system. The input values undergo mapping with the linguistic variables by the fuzzification and defuzzification process. The MF is termed as a curvature that will map every individual element to a collection of succeeding membership values in a closed interval [0, 1] which is depicted in Figure 2. A real crisp value is transformed to a fuzzy value using the MF. The arbitrary curve of the fuzzy MF should range between 0 and 1 so as to enhance the speed as well as the efficiency of the system.

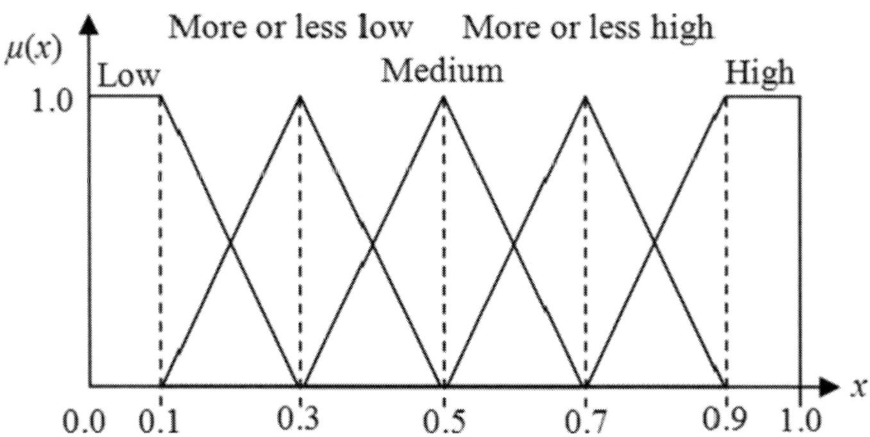

Figure 2. MF.

The simple MFs are generated by the use of straight lines, triangular MF (trimf) and trapezoidal MF (tramf). The triangular MF is produced by a total of three points whereas the trapezoid is constructed as a condensed triangle and has a smooth crest. An expert system can be determined as a universal Boolean valued information system comprising a non-empty determinate set of examples and non-empty determinate set of attributes. When (E, e)a soft set is over a universal set U, then (E, e) comprises of a Boolean Value.

Table 2. The input values of some patients

P	HbA1c	PGF	PGPP	AGE
P3	100	75,125	121,175	57
P19	67	85,110	135,180	68
P25	73	90,111	110,167	71
P42	82	88,105	119,170	59
P46	91	113	165	62
P55	78	109,135	152,229	70
P60	86	101,214	155,235	78
P68	93	120,226	162,270	66
P72	84	123,259	157,283	59
P75	97	117,321	151,215	79

Fuzzified Dataset

Fuzzy sets have a grade of MF. Fuzzy sets are widely used in areas like semantics, policy-construction and clustering where inaccurate concepts are frequently present. Consider a Universal set 'U', a set of parameters 'P' and a subset of 'X'. The mapped function F: U I [0,1] defines the fuzzy set F of U. As the dataset is not suitable for the application to soft sets directly as given in Table 2, the fuzzy membership values of the datasets specified in Table 3 are obtained by the formulas recommended by the expert doctors. The linguistic variables for HbA1 c, PG-F, PG-PP and age are extracted as very low (VL), low (L), middle (M), high (H), very high (VH) and (for age) young (Y), middle (M), old (O). These MFs are represented in equations (2), (3), (4).

$$H_b A_1 c(a) = \begin{cases} \mu_a \; if \; 5.7 < a < 7.0 \\ \quad 1 \; if \; 8.1 \leq a \end{cases} \tag{2}$$

$$PGF(b) = \begin{cases} \mu_b \; if \; 70 < b < 99 \\ 1 \; if \; 86 \geq 120 \end{cases} \tag{3}$$

$$PGPP(c) = \begin{cases} \mu_c \; if \; 140 < b < 200 \\ 1 \; if \; c \geq 120 \end{cases} \tag{4}$$

Conversion of Fuzzy Sets to Superlative Sets

Let us assume 'U' as a universal set and a group of parameters as 'P'. A soft set over the superlative universe (U,P) is defined by (F, X) where $F\!: X$R(U) and 'X' is a subgroup of P and $R(U)$ is the rule set of U. Every discrete fuzzy set can be precise as a soft set. The collection of parameters is designated by the usage of MFs. So, the whole collection of parameters holds numerals which are attained by relating the fuzzy set to soft set as indicated below and given in Table 3.

HbA1c:

P = {0 = P1, P2, P3... P78} E = {0,0.225,0.5,0.275,1}

(FM HbA1c, E) = {0 = {P4, P5, P6, P11, P13, P15, P20, P22, P23, P25,P30, P32, P34,P38, P41, P42, P43, P44, P53, P60, P73, P75},

0.225 = {P4, P6, P11, P13, P15, P20, P22, P23, P25, P34, P38, P41, P43, P44, P60, P64, P75},

0.5 = {P34, P38, P41, P43, P44}

0.275 = {P13, P20, P23, P38, P41},

1 = {P20, P38}}

PG-F,PG-PP:

P = {P1,P2,P3...P78}   E = {0, 0.175, 0.327, 0.525, 0.734}
(FB PG, E) = {0 = {P11, P17, P35, P36, P45, P46, P49, P53, P55, P60, P65, P68, P72, P73, P75},

0.175 = {P11, P17, P36, P45, P60, P68, P72, P73, P75},

0.327 = {P36, P45, P60, P68, P72, P73, P75},

0.525 = {P45, P68, P72, P75},

0.734 = Ø}}

Age:

P = {P1, P2, P3... P78} E = {0.08, 0.37, 0.59, 0.88, 0.97} (FM Age,E) = {0.08 = {P3, P8, P9, P22, P33, P35, P42, P43, P44, P46, P48, P49, P52, P56, P63, P66, P67, P69, P70, P72, P74, P76, P78},

0.37 = {P3, P22, P33, P35, P42, P43, P44, P48, P52,P58, P63, P66, P69, P70, P72, P74, P76, P78},

0.59 = {P43, P48, P52, P58, P63, P70, P74, P78},

0.88 = {P48, P52, P70},

0.97 = Ø}}

Reducing the Soft Set Parameters

Decreasing the parameters is to produce a set of rules by eliminating the redundancy and inconsistency. The complex resemblances among the rule sets are also reduced. A significant value is allocated to every rule in the system by the experts for creating an accurate degree of association among the attributes of the rule set. At this point, soft set represents every fuzzy set as attained below:

Table 3. The fuzzy membership values of some patients

P	HbA1c	PGF	PGPP	AGE
P3	1 VH	0.53S,0.47M	0.12S,0.17 M	0.48M, 0.540
P19	0.3 VL, 0.9L	0.77S,0.23M	0.13S,0.18M	1O
P25	0.51L, 0.53 M	0.8S,0.2M	0.11S,0.16M	1O
P42	0.31VL, 0.73 L	0.4S,0.6M	0.11S,0.17M	0.34M, 0.68O
P46	0.85VL, 0.73 L	1M	0.16M	0.14M,0.88O
P55	0.87VL, 0.17 L	0.93M, 0.07B	0.15M,0.22B	1O
P60	0.7L, 0.5L	0.6 M,0.4B	0.15M,0.23B	1O
P68	0.42L, 0.61 M	0.4 M,0.6B	0.16M,0.27B	1O
P72	0.67VL, 0.35 L	0.27M, 0.73B	0.15M,0.28B	0.34M,0.68O
P75	0.37L, 0.65 M	0.37M, 0.63B	0.15M,0.21B	1O

HbA1c:

P = {P1, P2, P3... P78} E = {0.225, 0.5, 0.275, 1}

(FM HbA1c, E) = {0.225 = {P4, P6, P11, P13, P15, P20, P22, P23, P25, P34, P38, P41, P43, P44, P60, P64, P75},

0.5 = {P13, P20, P23, P38, P41},

0.275 = {P13, P20, P23, P38, P41},

1 = {P20, P38}}

PG-F, PG-PP:

P = {P1, P2, P3... P78} E = {0.175, 0.327, 0.525, 0.734} (FB PG, E) = {0.175 = {P11, P17, P36, P45, P60, P68, P72, P73, P75},

0.327 = {P36, P45, P60, P68, P72, P73, P75},

0.525 = {P45, P68, P72, P75}}

Age:

P = {P1, P2, P3… P78} E = {0.37, 0.59, 0.88}

(FM Age,E) = 0.37 = {P3, P22, P33, P35, P42, P43, P44, P48, P52, P58, P63, P66, P69, P70, P72, P74, P76, P78},

0.59 = {P43, P48, P52, P58, P63, P70, P74, P78},

0.88 = {P48, P52, P70}}

Soft Rules Generation

Association rules determine the degree of relationship between the patients and the diagnosis report with the help of random sampling. Since each and every member of the set has the equal probability of getting chosen, random sampling is used to eliminate the division of the entire set into subsets. For example, consider a set of '*P*' objects and trial of '*N*' objects. When arbitrary sampling is employed, every probable trial of 'N' objects is expected to occur. A rule is arbitrarily created and computed totally by chance. This specific rule created is no way related to the other set of rules generated. The soft rules are attained using 'AND' operation as indicated below.

FVL HbA1c (0.35) $\wedge$ FM PG (0.25) $\wedge$ Fo Age (0.59) = {P7, P10, P14, P16, P27, P31, P33, P39, P46, P49, P50, P51, P54, P55, P56, P57, P61, P62, P63, P65, P67, P71, P72}

FL HbA1c (0.2875) $\wedge$ FS PG (0.275) $\wedge$ FM Age (0.31) = {P22, P33, P42, P43, P44, P48, P52, P58, P63, P70, P74, P78}

FM HbA1c (0.25) $\wedge$ FM PG (0.25) $\wedge$ Fo Age (0.325) = {P6, P11, P15, P20, P34, P41, P44, P60}

FM HbA1c (0.25) $\wedge$ FM PG (0.5) $\wedge$ Fo Age (0.325) = {P6, P11, P15, P34, P41, P44, P60}

FH HbA1c (0.2225) ∧ FS PG (0.785) ∧ Fo Age (0.59) = {P8} FH HbA1c (0.2225) ∧ FS PG (0.53) ∧ Fo Age (0.325) = {P5, P8}

FVH HbA1c (0.6875) ∧ FS PG (0.785) ∧ Fo Age (0.59) = { P8} FVH HbA1c (1) ∧ FS PG (0.275) ∧ Fo Age (0.59) = {P5, P8, P34}

Here, a set of around 658 rules is generated. Next, the rules which have identical set of output have been eliminated and the rule set is reduced to 435 rules.

Investigation Superlative Rules

The rules of associativity primarily hinge on the knowledge of support and confidence. Support is a measure of the degree of recurrence of an object which ensues in the dataset whereas Confidence governs the accurate value of the if and then rules. Rules are implicit in a production system. The syntax is represented as

If < condition > then < action > (5)

The rule defines that the action takes place upon the satisfaction of the condition. The rule-by-rule process supported by the conventional systems has been fixed by the classification model through corresponding rule stimulation. At this point, the soft rules are investigated and the percentage of diabetic risk is determined. Every individual rule is created for the respective set of patients. In accordance to these sets, the patient counts with the presence of diabetes are determined and the risk percentage is computed for each rule. When the patient data emerges more than once then the one with maximum risk percentage is considered.

PERFORMANCE EVALUATION

The presented Superlative Set based model is validated under several aspects and the results are explained below. Table 4 shows the sample fuzzy membership values of patients with respect to the varying number of attributes. At the same time, Table 5 provides the sample input values of patients. Table 6 and Figure 3 show the comparative analysis of different mathematical models under a set of 8 rules. The first two rules comes under the expert opinion of ≥ 30, the third rule comes under the opinion of ≥ 70, the fourth rule comes under ≥ 80 and the last four rules comes under ≥ 90.The grey prediction, rough set prediction and the fuzzy set theory have been used for comparing the performance of the proposed soft set model. Under the rule 1, the present model achieves a maximum of 28% whereas the grey prediction, rough set theory and fuzzy set theory attain a minimum of 15%, 13% and 12% respectively.

Table 4. Sample Fuzzy membership values of patients w.r.t attributes

P	HbA1c	PGF	PGPP	Age
P3	8.5VH	75S,125M	121S,175M	0.47M,0.530
P19	4.2VL,5.1L	85S,110M	135S,180M	650
P25	5.3L,7.1M	90S,111M	110S,167M	670
P42	4.7VL,5.3L	88S,105M	119S,170M	33M,670
P46	4VL,6.1L	113M	165M	30M,870
P55	4.7VL,5.5L	109M,135B	152M,229B	730
P60	5.4L,7.5M	101M,214B	155M,235B	680
P68	4.4VL,5.2L	120M,226B	162M,270B	750
P72	4.9VL,6.5L	123M,259B	157M,283B	33M,690
P75	6.7L,8.0M	117M,321B	151M,215B	830

Similarly, under the rule 2, the present model achieves a maximum of 34% whereas the grey prediction, rough set theory and fuzzy set theory attained minimum values of 25%, 27% and 23% respectively. Likewise, under the rule 3, the presented model achieves a maximum of 74% whereas the grey prediction, rough set theory and fuzzy set theory attained a minimum of 67%, 62% and 56% respectively. At the same time, under the rule 4, the present model exhibits superior results with a maximum of 83% whereas the grey prediction, rough set theory and fuzzy set theory attained a minimum of 64%, 58% and 61% respectively.

Table 5. Sample input values of patients

P	HbA1c	PGF	PGPP	AGE
P3	100	75,125	121,175	57
P19	67	85,110	135,180	68
P25	73	90,111	110,167	71
P42	82	88,105	119,170	59
P46	91	113	165	62
P55	78	109,135	152,229	70
P60	86	101,214	155,235	78
P68	93	120,226	162,270	66
P72	84	123,259	157,283	59
P75	97	117,321	151,215	79

Under the rule 5, the presented model achieves a maximum of 100% whereas the grey prediction, rough set theory and fuzzy set theory attained minimum values of 83%, 71% and 77% respectively. Similarly, under the rule 6, the presented model achieves a maximum of 100% whereas the grey prediction, rough set theory and fuzzy set theory attained minimum values of 75%, 59% and 69% respectively.

Table 6. Diabetic risk percentage of mathematical models w.r.t experts opinion

Sample Rules	Experts Opinion	Grey Prediction	Rough Set Theory	Fuzzy Set Theory	Superlative Soft Set Theory
Rule 1	≥ 30	15	13	12	28
Rule 2	≥ 30	25	27	23	34
Rule 3	≥ 70	67	62	56	74
Rule 4	≥ 80	64	58	61	83
Rule 5	≥ 90	83	71	77	100
Rule 6	≥ 90	75	59	69	100
Rule 7	≥ 90	92	82	85	100
Rule 8	≥ 90	96	95	83	100

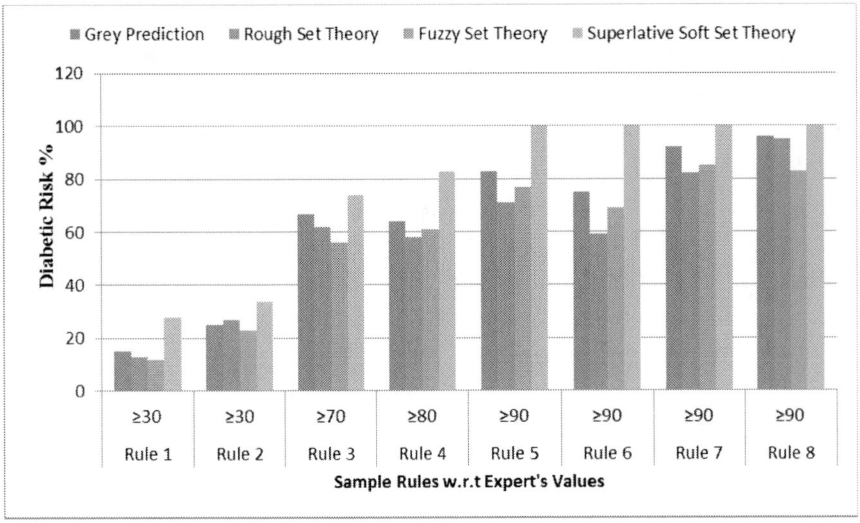

Figure 3. Diabetic Risk Percentage of mathematical models w.r.t Experts opinion.

Table 6 and Figure 3 show a comparative analysis of different mathematical models under a sample set of 8 rules. The first two rules come under the expert opinion of ≥ 30, the third rule comes under the opinion of ≥ 70, the fourth rule comes under ≥ 80 and the last four rules come under ≥ 90.

For comparison purposes, grey prediction, rough set prediction, fuzzy set theory and proposed model. Under the rule 1, the presented model achieves a maximum of 28% whereas the grey prediction, rough set theory and fuzzy set theory attain minimum values of of 15%, 13% and 12% respectively.

CONCLUSION

In this chapter, a Superlative Set based mathematical model is devised to estimate the diabetic risk percentage. The percentage of diabetic risk of every rule is calculated by providing a set of suppressed rules as input to the soft expert system. The rule sets are not in the specified format for applying them directly to generate a set of rules. Hence, the dataset is fuzzified and the corresponding membership values are obtained using the formulae given by the expert mathematicians. The real time dataset comprising the values of Gycosylated Haemoglobin, Plamsa Glucose- Fasting and Post Prandial values and age of the diabetic patients were collected from a multispeciality diabetic centre for analyzing the Superlative expert system. The set of diabetic values were given as input into the Superlative expert system and the results were determined as a percentage of diabetic risk generating a set of soft rules. The results were compared with the other existing mathematical models like the fuzzy set theory, grey prediction model and the Superlative soft set model has significantly achieved better performance in determining the percentage of diabetic risk.

References

[1] Lukmantoand R. B., E.Irwansyah, "The Early Detection of Diabetes Mellitus (DM) Using Fuzzy Hierarchical Model," *Procedia Computer Science*, vol. 59, no. 1, pp. 312-319, 2015.

[2] Polat K. and S. Güneş, "An expert system approach based on principal component analysis and adaptive neuro-fuzzy inference system to diagnosis of diabetes disease," *Digital Signal Processing*, vol. 17, no. 4, pp. 702-710, 2007.

[3] Gajare S. and S. Sonawani,"Improved Automatic Feature Selection Approach for Health Risk Prediction." *2018 Second International Conference on Computing Methodologies and Communication (ICCMC),* vol. 1, no. 1, pp. 23-42, 2018.

[4] Bellazzi R. and B. Zupan, "Predictive data mining in clinical medicine: current issues and guidelines," *International journal of medical informatics*, vol. 77, no. 2, pp.81-97, 2008.

[5] Lakshmi K., "Utilization of Data Mining Techniques for Prediction and Diagnosis of Tuberculosis Disease Survivability," *International Journal of Modern Education and Computer Science*, vol. 5, no. 8, pp. 8-17, 2013.

[6] Sanakal R. and S. T. Jayakumari, "Prognosis of Diabetes Using Data mining Approach-Fuzzy C Means Clustering and Support

Vector Machine," *International Journal of Computer Trends and Technology*, vol. 11, no. 2, pp. 94-98, 2014.

[7] Wishart D. S., "Emerging applications of metabolomics in drug discovery and precision medicine," *Nature reviews Drug discovery*, vol. 15, no. 7, pp. 473, 2016.

[8] Hood L. and M. Flores, "A personal view on systems medicine and the emergence of proactive P4 medicine: predictive, preventive, personalized and participatory," *New biotechnology*, vol. 29, no. 6, pp.613-624, 2012.

[9] Flores M., G. Glusman, K. Brogaard, N. D. Price, and L. Hood, "P4 medicine: how systems medicine will transform the healthcare sector and society," *Personalized medicine*, vol. *10, no.*6, pp.565-576, 2013.

[10] Bates D. W., S.Saria, L. Ohno-Machado, A. Shah, and G.Escobar, "Big data in health care: using analytics to identify and manage high-risk and high-cost patients," *Health Affairs*, vol. 33, no. 7, pp.1123-1131, 2014.

[11] Polit D. F., and C. T. Beck, *Nursing research: Generating and assessing evidence for nursing practice,* Lippincott Williams & Wilkins, 2008.

[12] Care P., "A clinical practice guideline for treating tobacco use and dependence," *JAMA*, vol. *283*, pp.3244-3254, 2000.

[13] Chen M., S. Mao, and Y. Liu, "Big data: A survey," *Mobile networks and applications*, vol.19, no.2, pp.171-209, 2014.

[14] Follen M., R. Castaneda, M. Mikelson, D. Johnson, A. Wilson, and K. Higuchi, "Implementing health information technology to improve the process of health care delivery: a case study," *Disease Management*, vol. *10, no.* 4, pp.208-215, 2007.

[15] Timmermans S. and M. Berg, *The gold standard: The challenge of evidence-based medicine and standardization in health care*, Temple University Press, 2010.

In: Research Challenges in Science ... ISBN: 978-1-68507-008-3
Editors: S. Kannadhasan et al.

Chapter 12

Investigation of Boost Converter Using Enhanced Dynamic Power Factor Rectification

R. Felshiya Rajakumari[1], S. Gopinath[2] and S. Pragadeswaran[3,*]

[1]Department of Electrical and Electronics Engineering, Karpagam Institute of Technology, Coimbatore, India
[2]Department of Electronics and Communication Engineering, Karpagam Institute of Technology, Coimbatore, India
[3]Department of Electronics and Communication Engineering, Karpagam Institute of Technology, Coimbatore, India

Abstract

In this work, boost converter power factor correction is analyzed. This work is discussed and design with a DC-DC boost converter and a single phase rectifier with improved power factor boost converter technique. The normal DC-DC boost converter is intend and the parameters such as ripple, input current, output voltage is executed and then power factor is calculated by

* Corresponding Author's E-mail: praga1994@gmail.com.

interfacing the boost converter with rectifier. By designing the input technique the power factor will be improved. Proactive power factor correction and active power factor correction for the boost converter is discussed in this paper. By using two power factor techniques active power factor correction which gives the best result because the boost converter is coupled with rectifier circuit. Boost converter is one of the best methods to implement for the power factor value from the input sinusoidal AC voltage waveform.

Keywords: converter, rectifier, ripple, power factor correction, total harmonic distortion

1. Introduction

In recent years switched mode power supply is used rapidly. SMPS are used as AC to DC sources for electronics appliances. The main disadvantage of using SMPS is draws a pulsating current and it has low power factor value due to high RMS value. So we are using active power factor and passive power factor correction technique method in this paper. A passive power factor component deals with resistance, inductance and capacitance, so it leads to poor power factor. So in this paper active power factor correction (ACPF) for DC-DC converters is very economical and efficient. The method of low power factor can be improved by using APFC. PFC deals with high frequency and it draws a sinusoidal AC current which is in phase with AC input voltage. Power factor correction is a good method in power system to achieve a good power factor value [1]. Advantages of boost converter other than DC-DC converter are that the buck converter output is always less than the input and it requires more additional component of passive filter and output of buck converter is discontinuous and input current does not follow input voltage. When boost and sepic converter are operated the output of both converters is discontinuous with high ripple voltage [2]. Comparing with buck, sepic converters, boost converter gives continuous input and output current with low ripple value and is best suited for power factor correction value.

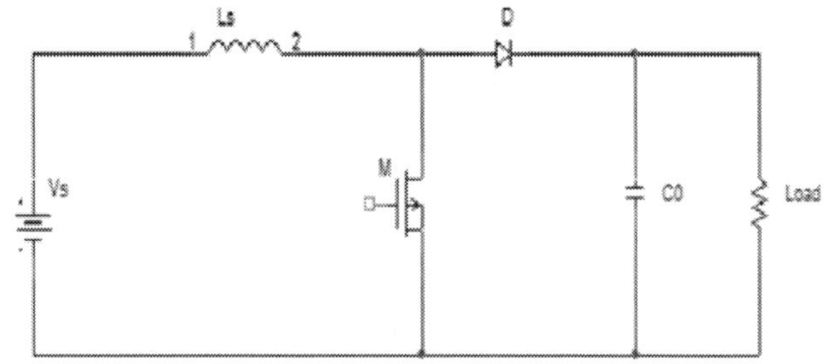

Figure 1. Schematic Diagram of Boost Converter.

Section I deals with the introduction of converter and SMPS. Section II presents the boost converter analysis operation.Section IIIprovides the power loss analysis performance calculation of boost converter. Section IV works with the boost converter power factor correction circuit. Finally conclusion is discussed in Section V.

2. Boost Converter

A boost converter provides an output voltage larger when compare with other converters.Figure 1 shows the schematic diagram of Boost converter. The diverse modes of operation of boost converter, design equations of the boost converter values are mentioned. The converter operates in diverse methods.

2.1. Method 1

The method 1 graphic diagram operation is shownin Figure 2.First operation starts when the controller (S_1) is turned on at$t=t_{on}$. When controller operation takes place, the inductor current rises linearly than zero and the current increases upward [3] [4]. During this operation, the output current is flow from the output capacitor for the entirely switch on-time.

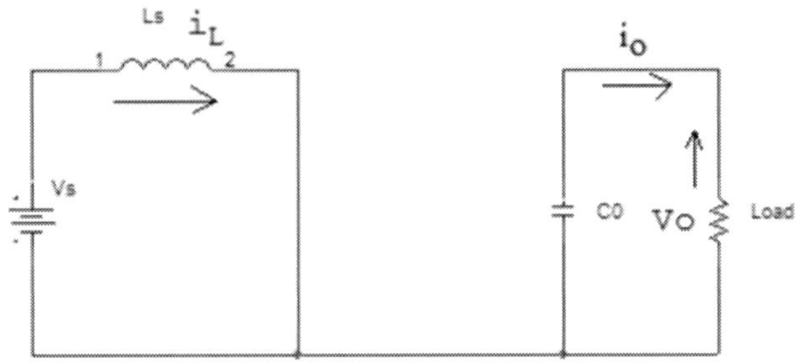

Figure 2. Mode 1 Schematic Operation of Boost Converter.

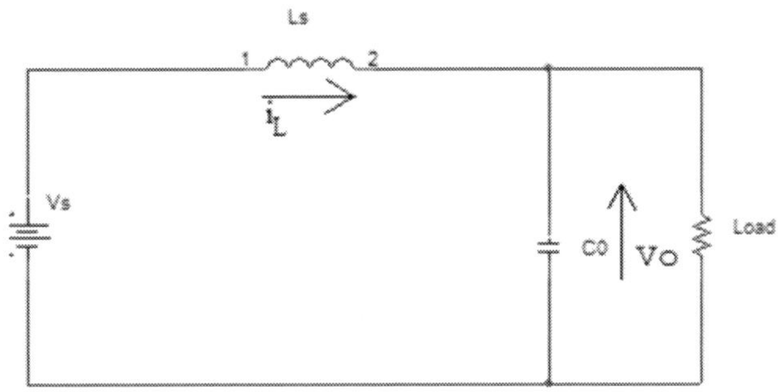

Figure 3. Mode 2 Schematic Operation of Boost Converter.

2.2. Mode 2

The method 2 circuit operation is shown in Figure 3. Second operation starts when the controller is turned off att = t_{on}. When come to regulator off operation, the inductor current diminishes rapidly and therefore the current circulates through R, L and C. In this operation, the inductor delivers the storedenergytothe load [3] [4].

2.3. Boost Converter Equations [5]

The conversion gain for the boost converter is given inEq. (1)

$$V_0 = \frac{V_s}{1-D} \tag{1}$$

Where V_s the source voltage, D is the duty cycle for the boost converter, V_0 is the output voltage.

The peak to peak inductor current ripple is given as inEq. (2)

$$\Delta I = \frac{V_s D}{f_s L} \tag{2}$$

$$L = \frac{V_s D}{f_s L} \tag{3}$$

where, L is the inductor.

The peak to peak capacitor voltage ripple is given byEq. (4)

$$\Delta V_c = \frac{I_a D}{f_s C} \tag{4}$$

$$\Rightarrow C = \frac{I_a D}{f_s \Delta V_c} \tag{5}$$

where, f_s is the switching frequency, C is the Capacitor, ΔV_c is the peak to peak capacitor voltage.

3. Power Loss Analysis Calculation for Boost Converter

The analysis calculation of the boost converter includes MOSFET loss, diode loss used in the converter circuit. The switching loss, conduction loss, is acquired.The power loss analysis of the converter includes the power loss of the MOSFET and diode. The conduction losses and switching losses are calculated and the results are obtained.

3.1. Power Loss of Mosfet

The power loss of MOSFET consists of the switching loss and the conduction loss.

$$P_{mosfet} = P_{switching} + P_{conduction} \tag{5}$$

$$P_{cond(mosfet)} = I^2 drain(rms) * RDS_{(on)} \tag{6}$$

$$P_{switching} = 0.5 V_{DS} * I_{drain} * f_{sw} * (t_{on} + t_{off}) \tag{7}$$

where,

I_{drain} = drain current of the MOSFET

f_{sw} = switching frequency

t_{on} = ON-period of the switching MOSFET

t_{off} = OFF-period of the switching MOSFET

$RDS_{(on)}$ = ON-state drain source resistance of MOSFET.

3.2. Calculation for Power Loss of Mosfet

$$P_{cond(mosfet)} = I^2 drain(rms) * RDS_{(on)} \tag{8}$$

3.3. Power Loss of Diode

$$P_{loss\_diode} = I_{rms\_diode} * V_f \tag{9}$$

$$I_{rms\_diode} = \frac{\Delta i * \sqrt{(1-D)}}{\sqrt{3}} + I_{min} * \sqrt{1-D} \tag{10}$$

$$I_{min} = \frac{I_{out} + \Delta i * \frac{(D-1)}{2}}{1-D} \tag{11}$$

$$\Delta i = \frac{DV_{in}}{Lf_{sw}} \tag{12}$$

where,

V_f = forward voltage drop
D= Duty cycle
L= Inductor
I_{min} = minimum current of the diode
I_{out} = output current of the BC.

4. Power Factor Correction for Boost Converter

Owing to the non-linear relationship of voltage and current wave form, the several converters produce harmonics. So it leads to low harmonic distortion, low poor factor. Poor power factor will leads to high RMS value of current. Now the boost converter will border with rectifier circuit and can be found out some of the important parameters such as power, total harmonic distortion for the research work process. . The main objective of power factor correction is to make input power supply behave like a resistive load [8].

Passive power factor correction is used to improve the powerfactor for the non linear load such as resistors, capacitors and inductors. Active power factor correction is based on switch mode converter. By this technique we can find out the important parameters such as total harmonic distortion and power factor [9, 10]. By using this active power factor correction technique, power factor will be improved in Boost Converter.

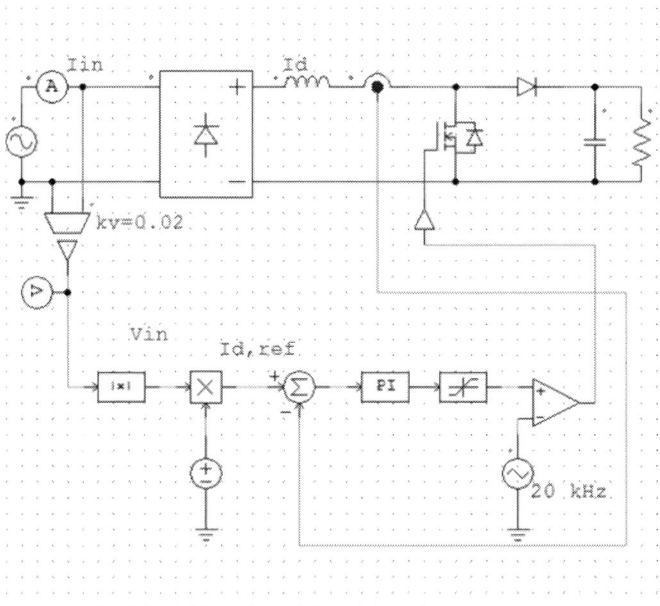

Figure 4. Schematic diagram for boost converter power correction.

Figure 4 shows the graphic diagram for boost converter powercorrection circuit diagram. It consists of three sections such asrectifier, boost converter and control circuit. The output of boost converter is given to error amplifier where it is compared with a reference voltage and the error voltage is amplified and it is fed to a multiplier. The switching pulse is produced from the comparator and it isused to trigger the pulse generator.

CONCLUSION

This work discussed with the DC-D C boost converter for finding the best power factor correction. First boost converter graphic plan is carried out in PSIM software. Power Loss calculation formula is discussed in the boost converter suchas power Mosfet, power diode loss. Finally boost converter is interface with rectifier and power factor correction is discussed by using active power factor correction techniques. By using APFC method, it gives the best power factor value and it brought closer to unity value.

REFERENCES

[1] Farzad Mohammadzadesh Shahir, et al., "Extended Topology of boost DC-DC Converter," *IEEE Transactions on Power Electronics,* Vol.34, Issue: 3,March2019.

[2] Preethi Sontakki, A.D.Srinivasan, "Sepic Topology Based High Voltage Gain DC-DC converter," *International Journal of Recent Technology and Engineering*, ISSN:2277-3878, Vol.8,Issue:1, May2019.

[3] S. Gopinath, S. Pragadeswaran, etal., "Assessment Work on Blue Brain Technology", *International Journal of Advanced Engineering Science and Information Technology (IJAESIT)*, Volume 5 and Issue V, Pages 35-38, ISSN: 2349-3216, May 2021.

[4] Shafinaz A. Lopa, Shahzad Hossain, et al., *Design and Simulation of DC-DC converters,* January,2018.

[5] Sandeep Tyagi, Garmia Verma, "Simulation and Analysis of DC-DC Boost Converter Using Sliding Mode Controller under Variable Conditions," *IOSR Journal of Electrical and Electronics Engineering,* ISSN:2278-1676,p-ISSN:2320-3331, Vol.13, Issue:1, Jan-Feb. 2018.pp.33-41.

[6] Gayathri S, Pragadeswaran S, "Landmine Detection using Drone" *International Journal of Advanced Engineering Science and Information Technology (IJAESIT)*, Volume 5 and Issue V, Pages 49-59, ISSN: 2349-3216, May 2021.

[7] Vineeth V., Arun Uday, et al., "Power Factor Correction for Boost Converter," *International Journal of Science and Research,* ISSN: 239-7064, Volume 6,Issue 5,May2017.

[8] Shafinaz A. Lopa, S. Hossain, "Design and Simulation of DC-DC converters," *International Research Journal of Engineering and Technology*, Vol. 3, issue:1, Jan2016.

[9] S. Pragadeswaran, S. Gayathri, et al., "Wireless Sensor Networks Security Issues, Security Needs and Different Types of Attacks based on Layers: A Survey" *International Journal of Advanced Engineering Science and Information Technology (IJAESIT)*, Volume 5 and Issue V, Pages 60-67, ISSN: 2349-3216, May 2021.

[10] Hiten Pahilwani, "Power Fcator Correction Using Boost Converter," *International Journal of Application or Innovation in Engineering and Management,* ISSN: 2319-4847, Volume 5, Issue 8, August 2015.

[11] P.V.R.K.B.A.N.Raju, Sudhakar Babul. et al., "Simulation of Active Power Factor Correction Using Boost Type Converter," *International journal of Science, Engineering and Techology Research*, Volume 3, Issue10, October2014.

[12] P. Yohan Babu, P. Surendra Babu, et al., "Power Factor Imrovement with High Efficiency Converters," *IOSR journal of Engineering,* Volume 25, pp: 1000-1006,May2012.

[13] S. Kannadhasan and R. Nagarajan, Development of an H-Shaped Antenna with FR4 for 1-10GHz Wireless Communications, *Textile Research Journal*, DOI:10.1177/00405175211003167 journals. sagepub.com/home/trj, March 21, 2021.

[14] S. Kannadhasan and R.Suresh, “EMD Algorithm for Robust Image Watermarking” *Advanced Materials Research* Vols. 984-985 (2014) PP 1255-1260, ISSN No:1022-6680.

[15] S. Kannadhasan and R. Nagarajan, Performance Improvement of H-Shaped Antenna With Zener Diode for Textile Applications, *The Journal of the Textile Institute*, Taylor & Francis Group, DOI: 10.1080/00405000.2021.1944523

[16] S.Kannadhasan, G.Karthikeyan, et al., “A Graph Theory Based Energy Efficient Clustering Techniques in Wireless Sensor Networks”. *Conference Proceedings by IEEE Explore Digital Library* 978-1-4673-5758-6/13 @2013 IEEE.

[17] Kannadhasan, M.Isavani, et al., “A Novel Approach Privacy Security Protocol Based SUPM Method in Near Field Communication Technology”, *Advances in Intelligent Systems and Computing*. 23-24 April 2014.

In: Research Challenges in Science ... ISBN: 978-1-68507-008-3
Editors: S. Kannadhasan et al.

Chapter 13

HYDROGEN FUEL CELL FOR AUTOMOBILE INDUSTRIES: A REVIEW

Nilay Poharkar[1], Ayushi Gupta[2,*], Akshay Srivastava[2], Jeet Patel[3] and Siddhi Vinayak Pandey[3]

[1]Department of Mechanical Engineering,
JSPM's Rajashri Shahu College of Engineering, Pune, India
[2]Department of Electronics and Communication Engineering,
Pranveer Singh Institute of Technology, Kanpur, India
[3]Department of Electrical Engineering,
Adani Institute of Infrastructure Engineering, Ahmedabad, India

ABSTRACT

Many automobile industries are moving towards the inclusion of renewable based energy generating resources due to the pollutions causes with nonrenewable (gasoline) based vehicles. Among them, electric and fuel cell based vehicles are leading in field of automobile industries Electric vehicles may not be an effective alternative for conventional gasoline conveyances due to

---

* Corresponding Author's E-mail: ayushiaggra2001@gmail.com.

its dependency nature on the way of electricity engenderment. The main aim of automakers is to minimize the exhaust emission and fuel consumption with optimizing the range, power and overall efficiency. In this paper, the rudimentary working principle of fuel cell along with its types, different kind of membranes and sundry comparisons such as - fuel cell electric vehicle with battery operated vehicle is elucidated in detail. This paper also present various kind of challenges and its solutions which required during the astute and efficient development of hydrogen fuel cell based infrastructure.

INTRODUCTION

The global energy demand has increased drastically due to the perpetual increment in the world's population [1]. Fossil fuel is the key essentiality to consummate this escalating energy demand which is having a significant pitfall [2]. The scarcity of reserves, dubious distribution of reserves, incrementing prices, early consumption of facilely accessible reserves and greenhouse gas emissions are some of the major drawbacks of fossil fuels consumptions [3]. Automobile industries are consuming huge amount of fossil fuel. This kind of industry also needs a transition change from traditional to smart and clean energy based resources [4]. The researchers have found hydrogen as a potential energy source which supersedes the traditional energy sources for future energy systems by the utilization of abundant and renewable energy predicated resources. The air combustion of hydrogen engenders energy and dihydrogen monoxide as a product; which can be engendered from renewable sources and will be felicitous for different kinds of applications such as specific heat production and centralized energy production. Apart from it, hydrogen has the competency to solve current environmental issues like climate changes and air pollution caused by exorbitant utilization of fossil fuels [5]. While, for fully commercialization of hydrogen technology there is a desideratum of detailed research and analysis to identify the pitfalls and advantages of hydrogen [6]. Hydrogen is the cleanest source of energy which has a three times higher heating value than petroleum. But, being an indirect source of energy, it needs to be processed before the authentic use which results in increase of the engenderment cost of hydrogen as an alternative fuel [7]. This alternative fuel is having a cost of more than three times compared to the initial cost of fossil fuels [8, 9]. In 1839 Sir William Grove started fuel cells, it was unsuccessful in the initial stages as electricity was unknown. While, the first favourable outcome was achieved by Francis Bacon in 1932. The attempt was achieved by using the setup of an alkaline fuel cell system using porous electrodes. Also, in the Apollo space program; the fuel cells were used in the year 1950. It was one of the finest resources of fuel for space programs as { the nuclear are treacherous, solar is very sizable, and batteries are ponderous to carry for space satellites. In more recent decades, a number of major automobile manufacturers and many government agencies have supported for the research and development of fuel cell technology for propelling hydrogen powered fuel cell vehicles [10]. The Proton exchange membrane fuel cell (PEMFC) suits best as it is having a many advantage like -low operating temperature, sustained operation at high current density, compact design, less weight, fast start-up, long life of stack and low costs. These factors make the use of PEMFC most promising for transportation purpose, which helps the efficient and astute workflow of hydrogen fuel cells [11].

Basic Working Principle of Fuel Cell

Electricity is generated in fuel cells by electrochemical reactions. The fuel cells have four major parts i.e., anode, cathode, electrolyte and external circuits [12]. Both the electrodes are made up of platinum (which acts as a catalyst during the process). The redox reaction takes place inside the fuel cell between the hydrogen and oxygen, this process also involves the conversion of chemical energy into the electrical energy.

Firstly the hydrogen is extracted (as it is not present in free state) from water using the electrolyser, then hydrogen is injected at the anode terminal of the fuel cell where it breaks into hydrogen ion, releasing the proton and electron with the reaction. The Proton gets attracted towards cathode (which has a supply of oxygen) whereas electron travel towards the cathode through load or external circuit generating electricity as a load. At the end of the process, hydrogen atom combines with the oxygen leaving behind water as the residue, later on which can be use as the coolant or can be reused. As mentioned in the chemical equations; the working mechanism for this application is explained. The chemical equation of Cathode and Anode is given as: $O_2 + 4\ H^+ + 4\ e^- \rightarrow 2H_2O$; $2H_2 \rightarrow 4\ H^+ + 4\ e^-${. While the overall equation is defined as: $2\ H_2 + O_2 \rightarrow 2H_2O$ [**?**]

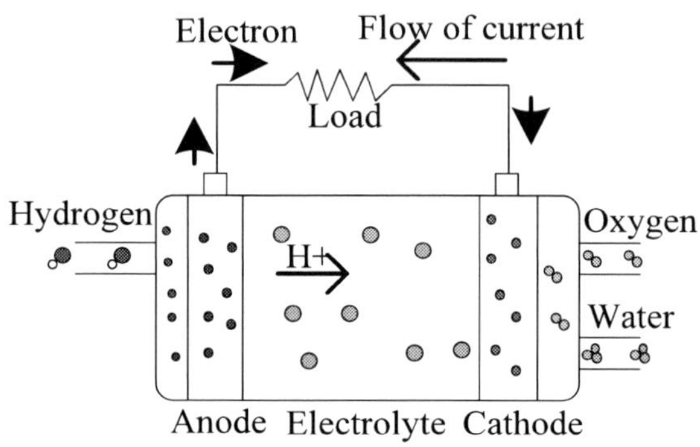

Figure 1. Internal structure of Fuel Cell.

There are basically six types of fuel cells are used on the basis of electrolytes –

- Alkaline fuel cell (AFC)
- Phosphoric acid fuel cell (PAFC)
- Solid oxide fuel cell (SOFC)
- Molten carbonate fuel cell (MCFC)
- Proton exchange membrane fuel cell (PEMFC)
- Direct methanol fuel cell (DMFC)

Alkali Fuel Cell

Discovery of Alkali Fuel Cell (AFC) begin in late 1930s, when Francis Thomas Bacon, a British Engineer started an experiment with alkali electrolyte. In that experiment he used potassium hydroxide instead of acid electrolyte and he got to know that potassium hydroxide are much more better than acid electrolyte as they do not corrode the electrode. The electrode he used in experiment were gas diffusion electrode which have pores in it. Also, he used the pressurized gases to keep the tiny pores of electrode flooded by the electrolyte but in late 1950 Union Carbide come up with the new idea of using carbon gas diffusion electrode. And further in early 1960 Eduard Justi, a German scientist launched a new gas diffusion electrode which were made up of nickel sponge on carbonyl nickel matrix [13, 14]. It uses potassium hydroxide (KOH) as the electrolyte, during the whole process hydroxyl ions are transferred from cathode to anode [15], it uses nickel alloys as catalyst. The Alkaline Fuel Cells (AFCs) operates at a temperature ranging from 60 to 90 degree Celsius. AFCs are double layer structure (Hydrophobic layer and Electrocatalyst layer) and they do not have bipolar plates. In AFCs, water and heat is removed as by product while recirculating the electrolyte.

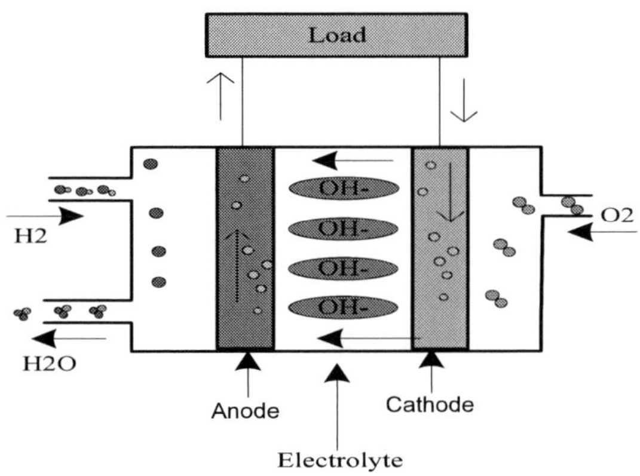

Figure 2. [17] Alkaline Fuel Cell.

Working

Firstly at anode, the hydrogen remains in zero oxidation state; while its resultant with water remains at one oxidation state. Which is clearly seen by the chemical equation of Cathode: $2H_2 + 4\ OH^- \rightarrow 4\ H_2O + 4e^-$. In case of the cathode, initially the oxygen remains at zero oxidation state; while the resultant oxygen in hydroxyl ion remains at two negative oxidation state. Therefore, the reduction process takes place within the process which is clearly explained by the Cathode equation: $2\ H_2 + 0_2$

→ 2 H_2O. By this equation, it is clearly visible that the hydrogen acts as a reducing agent; while the oxygen acts as an oxidizing agent.

Phosphoric Acid Fuel Cell

Discovery of Phosphoric Acid Fuel Cells (PAFC) ran on air in 1961 when two scientists G. V. Elmore and H. A. Tanner performed an experiment by using 35 percentage phosphoric acid and 65 percentage silica powder annihilated into a Teflon cord. In that experiment they noticed that percentage of phosphoric acid remained same under cell operating conditions, not like the sulphuric acid which was used by William Grove in his experiment in 1842. Further in mid-1960, army researchers of U.S. reconnoitered the potentials of PAFCs. Here also, carbon gasdiffusion layer discovered by Union Carbide incorporated with TARGET (Team to Advance Research foe Gas Energy Transformation, NC) was used. PAFCs operate at the temperature from 150° to 200°C [16], uses phosphoric acid as electrolyte and carbon paper as electrodes [13] like AFCs, PAFC also follow the same process of electrons and proton transfer due to which water and heat is formed . At atmospheric pressure, exploitation of water for steam generation causing formation of lot carbon monoxide which will further affect the performance of fuel cell. This issue can be resolved by increasing the temperature at the anode as per TARGET incorporation.

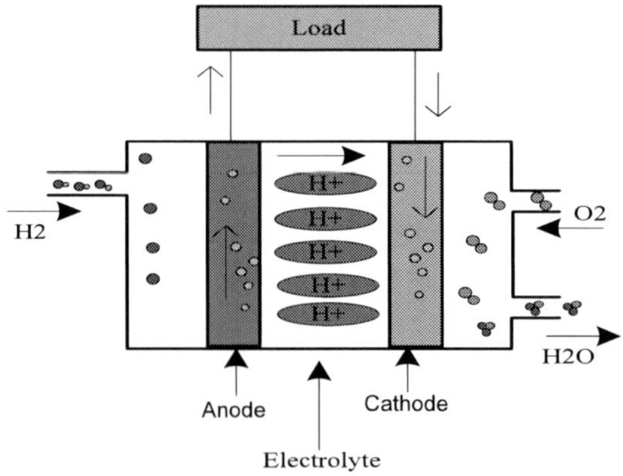

Figure 3. [16] Phosphoric Acid Fuel Cell.

Working

Firstly at anode, the hydrogen remains at zero oxidation state, while the resultant hydrogen remains at one oxidation state; which causes the oxidation of hydrogen at the anode side of the system. The chemical equation for this reaction is: $H_2 \rightarrow 2H^+ + 2e^-$. Similarly at cathode, initially the oxygen remains at zero oxidation state; while the resultant oxygen remains at negative two oxidation state. The chemical equation for this reaction is:

$$50_2 + 2H^+ + 2e^- \rightarrow H_2O.$$

Solid Oxide Fuel Cell

Discovery of Solid Oxide Fuel Cells (SOFC) was rooted in 1930 when two scientist Emil Baur and H. Preis from Switzerland were working on a project which was aimed to replace carbon filament lamps. They started working on solid oxide electrolytes like Zirconium, Yttrium, Cerium, Lanthanum, Tungsten etc. But the results were not good as expected so in 1940; O. K. Davtyan a Russian scientists tried by adding monazite sand to a mixture of soda glass, tungsten trioxide, sodium carbonate. This mixture was used to increase the mechanical strength and conductivity. And then in late 1950, the research in SOFCs tried to reach at zenith. But in 1959, researchers noticed that; when used in fuel cells, solid electrolytes had many problems like short-circuiting because of semi conductivity, high internal electrical-resistance, melting and many more. Then in 1962, it was noted that 40 companies are still working on SOFCs.

SOFC operates at temperature ranging from 500 - 1000°C and it is capable of conducting 0 2{ from cathode to anode. Since it operates at a very high temperature so the compounds should be capable to handle this much amount of temperature; which is used at the anode and cathode side of the system. The Electrolyte used is YSZ (Yttria Stabilized Zirconia); which is chemically and thermally stable and purely ionic in nature [17]. Their efficiency is upto 70% which can be further increased by the coagulation approach [18].

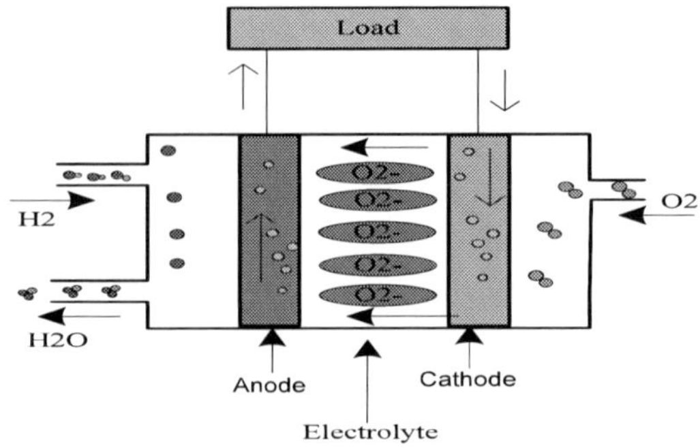

Figure 4. [21] Solid Oxide Fuel Cell.

Working

Firstly at anode, the hydrogen remains at zero oxidation state, while the resultant hydrogen remains at one oxidation state; which causes the oxidation of hydrogen at the anode side of the system. The chemical equation for this reaction is: $O_2 + H_2 \rightarrow H_2O + 2\ e^-$. Similarly at cathode, initially the oxygen remains at zero oxidation state; while the resultant oxygen remains at negative two oxidation state. The chemical equation for this reaction is: $.5O_2 + 2e^- \rightarrow O_2$. The overall equation of a process is defined as: $.5O_2 + H_2 \rightarrow H_20$; in this process the hydrogen acts as a reducing agent, while the oxygen acts as an oxidising agent for the system.

Molten Carbonate Fuel Cell

Discovery of Molten Carbonate Fuel Cell (MCFC) was started with the discovery of SOFC. When Emil Baur and H.Precis worked on solid oxide electrolyte and come up with the issues in that electrolyte; after that, in late 1950, two Dutch scientists G.H.J. Broers and J.A.A. Ketelaar worked on the limitations in solid oxide electrolyte and he got fused/molten carbonate salts as substitute. In 1960, they worked upon this fused carbonate electrolyte and found it as mixture of potassium carbonate, sodium carbonate and lithium carbonate soaked in a porous sintered platter of magnesium oxide. But, after few years scientists find that molted carbonate was not able to work fine with cord (gasket) materials.

MCFC operates at the temperature ranging from 600-650°C [22] and it conduct carbonate ion from cathode to anode. Due to its high operating temperature it uses hydrocarbon fuels that does not require any reforming. It uses molten carbonate mixture as electrolyte, non-precious metal can be used as electrodes as they operate at very high temperature. MCFC also operates at a very low current density.

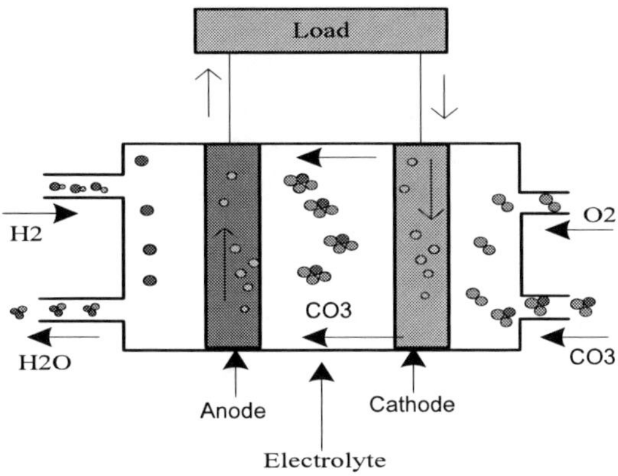

Figure 5. [19] Molten Carbonate Fuel Cell.

Working

Firstly at anode, the hydrogen remains at zero oxidation state, while the resultant hydrogen remains at one oxidation state; which causes the oxidation of hydrogen at the anode side of the system. The chemical equation for this reaction is: $H_2 + CO_3^{2-} \rightarrow H_2O + CO_2 + 2e^-$. Similarly at cathode, initially the oxygen remains at zero oxidation state; while the resultant oxygen remains system at negative two oxidation state. The chemical equation for this reaction is: $CO + CO_3^{2-} \rightarrow H_2O + CO_2 + 2\ e^-$. The overall equation of a process is defined as:.5 $O_2 + H_2 \rightarrow H_2O$; in this process the hydrogen acts as a reducing agent, while the oxygen acts as an oxidising agent for the system.

Proton Exchange Membrane Fuel Cell

After the extermination of MCFCs in early 1960, Thomas Grubb and Leonard Niedrach started experimenting with polymer electrolyte membranes (PEM) and they firstly got that it is able to work excellent with gasket materials. They used this product in a fuel cell for U.S Navy's Bureau and also for U.S Army and called it as Proton Exchange Membrane Fuel Cells (PEMFC). In that fuel cell he used platinum catalysts which was expensive, but it was compact and portable. It was also used for NASA's Space Program Gemini and then for many more projects. In mid-1970, scientist named GE developed proton exchange membrane (polymer electrolyte membrane) for water electrolysis which was used in major of the programs of U.S. Navy and NASA Programs. Now developers added GoreTex to PEM cells to increase the strength of electrolyte. This fuel cell is classified in two types; one operates at the temperature range of 60-80°C and other operate at the temperature range of 110-180°C; which is also known as the high temperature proton exchange membrane fuel cell. The PEMFC that operate at low temperature uses water-based electrolyte; while the high temperature uses the mineral-acid based electrolyte [24]. Also, the platinum-ruthenium is used as a catalyst for high temperature; while the simple platinum is used for the low temperature [19].

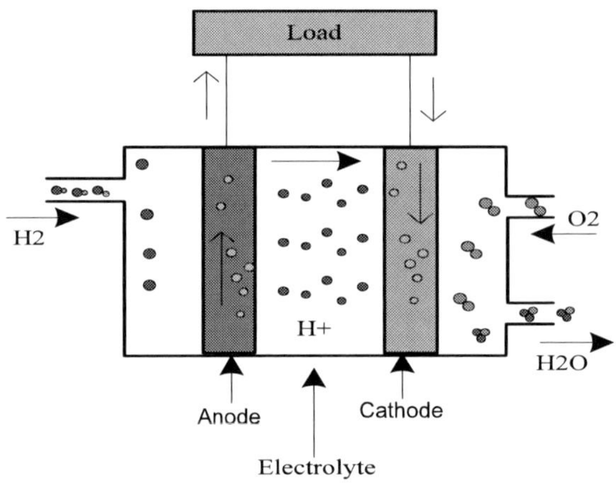

Figure 6. [20] Proton Exchange Membrane Fuel Cell.

Working

Firstly at anode, the hydrogen remains at zero oxidation state, while the resultant hydrogen remains at one oxidation state; which causes the oxidation of hydrogen at the anode side of the system. The chemical equation for this reaction is: $2\ H_2 \rightarrow 4\ H^+ + 4\ e^-$. Similarly at cathode, initially the oxygen remains at zero oxidation state; while the resultant oxygen remains at negative two oxidation state. The chemical equation for this reaction is: $O_2 + 4\ H^+ + 4e^- \rightarrow 2\ H_2O$. The overall equation of a process is defined as:

$$2\ H_2 + O_2 \rightarrow 2\ H_2O$$

In this process the hydrogen acts as a reducing agent, while the oxygen acts as an oxidising agent for the system.

Direct Methanol Fuel Cell

Direct Methanol Fuel Cell (DMFC) has been discovered after discovery of PEMFCs. They are the advanced version of PEMFCs. Among all other fuel cells, it does not need to be recharged and have a long life. It uses methanol as an energy source [17]; which is similar to the PEM, that uses polymer membraneas an electrolyte of a system. It operates at a temperature ranging of 40-90°C [21].

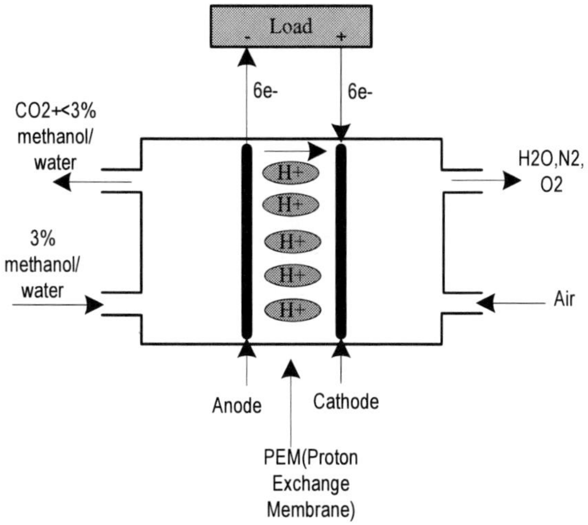

Figure 7. [22] Direct Methanol Fuel Cell.

Membrane

A thin sheet of material act as barrier or boundary is known as membrane. There are four types of ion exchange membranes are available; which are:

1. Cation Exchange Membrane (CEM)
2. Anion Exchange Membrane (AEM)
3. Bipolar Membrane (BPM)
4. Charge Mosaic Membrane (CMM)

On the basis of their electrolysis performance, the order is maintained such as: AEM > CEM > CMM > BPM; But when it comes to transportation of number of protons and ability to reduce increment in pH in the cathode chamber, they are ordered like: BPM > AEM > CMM > CEM [23]; So, for proton exchange membrane fuel cells to operate the electrolytic performance very well, Cation Exchange Membrane (CEM) is used as here it is required to transfer proton ions (H+) which are cations and also it does not allow anions to pass through it. So, Cation Exchange Membranes (CEM) are used as Proton Exchange Membrane (PEM) in PEMFCs.

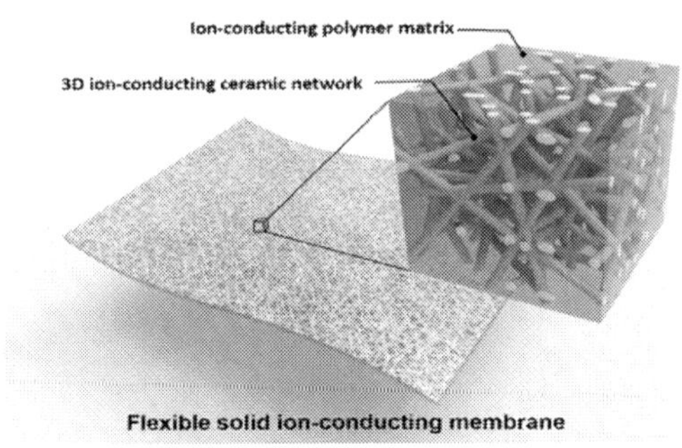

Figure 8. [25] Microscopic View of Ion Exchange Membrane.

Figure 9 shows the schematic diagram of the hybrid composite electrolyte solid in state. The nanofibers made up of ceramic garnet functions as the strength provider and ion-conducting polymers; which works as a maze or a matrix of the system. The Nanofibers form a tube-like 3D structure which are distributed randomly and interconnected for continuous ion conduction. The figure also depicts the structure of 3D Li7La3Zr2O12 polymer composite membrane. These 3D ceramic garnet-polymer composite membranes do not need to mix fillers with polymers mechanically [24].

Polymer Electrolytes

The Figure 2 shows a Polymer Electrolyte Membrane along with reactions which are occurring at each electrode and in this figure the orange spheres denote electronically conductive carbons. Here, two irreversible reactions are occurring at the two different electrodes. Although both the reactions are reversible, but fuel cells exploit only the forward reactions giving water as a residue to the atmosphere. The role of this type of polymer electrolyte is similar with the polymer electrolytes in the lithium battery; except for the ion used as a H+. The covalent bond is formed between anion and the polymer backbone in these membranes, and transport of H+ occurs in the absence of anion transport. Therefore, transportation of non-energy-producing ions causes no energy loss [26].

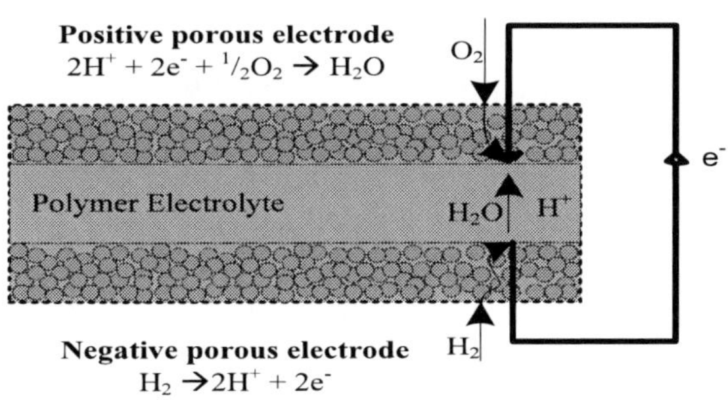

Figure 9. [26] Schematic diagram of Polymer Electrolyte Membrane.

Proton Exchange Membrane

The Proton Exchange Membrane or Polymer Electrolyte Membrane (PEM) is a thin semipermeable membrane or a thin sheet of ionomers which was designed to conduct protons (Hydrogen ions); and also acts as an electronic insulators and reactant barriers. Hydrogen is activated in Proton Exchange Membrane Fuel Cell (PEMFC) to form a proton ion and eject electrons at the anode. The electron is forced to pass through the external circuit and proton passes through the membrane; which generates the electricity. The electron then flows back to the cathode and interacts with proton ion and oxygen to form the water. The PEMFC also consists from the bipolar plates and membrane electrode assembly (MEA). The MEA is composed of gas layer diffusion, catalyst layer and the membrane which is composed between two porous carbon electrodes coated with a minimum amount of platinum catalyst. The most common material for the membrane; which is a Nafion, a perfluorinated sulphuric acid polymer. The transfer of proton from anode to cathode is achieved by the membrane and also blocks the passage of reactants and electrons. To access the uniformity of fuel gas; the diffusion layer is used. Electrons at anode passes through the external circuit and generate electricity. PEMFCs are compact and lightweight weight systems and the sealing of the electrodes in PEMFCs is easier than other fuel cells because of the solidity of the electrolyte. The higher efficiency can be gained at higher working temperatures due to high reaction rate. Electrical efficiency of PEMFC is 40-50%. The hydrogen is an interesting medium of storing excess renewable energy as it can be used in a variety of applications like power generation, for supplementation of natural gas grid for increased efficiency, chemical feedstock for eco-friendly generation of fertilizers, or vehicle fueling. It's membrane-based system which overcomes many of the disadvantages of alkaline liquid systems, as the carrier fluid is in the form of deionized water, and the membrane-based cells design provides differential pressure operation [26].

Advantages

The PEM can be operated in wide range of power input and has low gas cross permeation that will prevent the direct contact of oxygen with hydrogen reducing the risk of flammable [27]. It has higher current density because there is a differential pressure.

One of the major advantage of proton electrolyte membrane is to compress hydrogen without inflation and leakage, it heat up quickly which will further enhance the electrolysis process [28] and cool down time is fast enough that no more power is required to liquify the hydrogen making it less inflammable, the inflammation that is caused between the hydrogen and oxygen is prevented by the proton exchange membrane which helps to individually pressurize each other. Since proton transportation is very short as well as thin, hence ohmic loss is very low. Ohmic losses are caused by the temperature variation, lower temperature enhances the electronic conductivity and higher temperature enhances the ionic conductivity [29]. Ionic conductivity is more effective as it operates at higher temperature causing less ohmic loss [30].

Disadvantages

The PEM uses expensive materials like bipolar plates, current collectors, membranes, noble catalyst increasing its manufacturing cost. PEM are very sensitive to dust, imperfections, impurities, they get easily damaged when overheated.

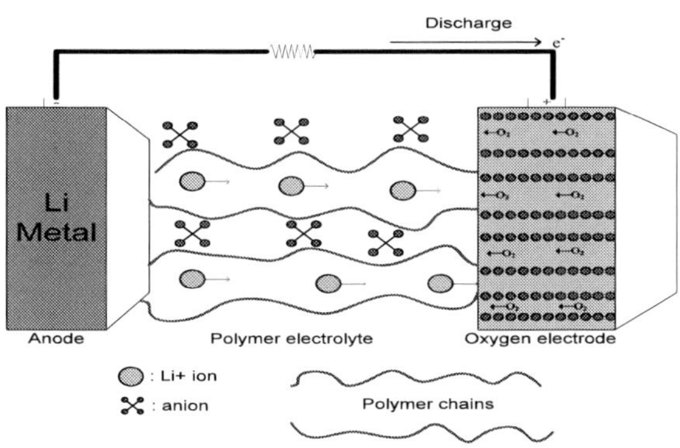

Figure 10. [31] Proton Exchange Membrane Fuel Cell (PEMFC).

COMPARISON OF FUEL CELL ELECTRIC VEHICLES WITH BATTERY ELECTRIC VEHICLES

Based on the Mass Density

The specific energy density in Wh/kg of Fuel Cell Electric Vehicles (FCEV) and Battery Electric Vehicles (BEV) are different. The specific energy density of BEV is compared with the types of batteries have been used within a conveyance, such as: Led {acid battery, Nickel-metal hydride battery and Lithium-ion battery used in Electric Vehicles determines the specific energy density of a vehicles. While, the hydrogen cell with different pressures such as: 35MPa and 70MPa are compared with the fiber wrapped composite tanks. As the pressure of the tank increases the utilization of fiber within a tank is also increases.

The energy plotted is the stored energy in form of electricity, used to power up the wheels. For batteries if the stored energy is of 200 KWh/Kg than available power will be nearabout 140*KWh*/Kg. This is because of the battery, which cannot be discharged below 20% of State of Charge (SOC) and cannot be charged above the 90% of its SOC.

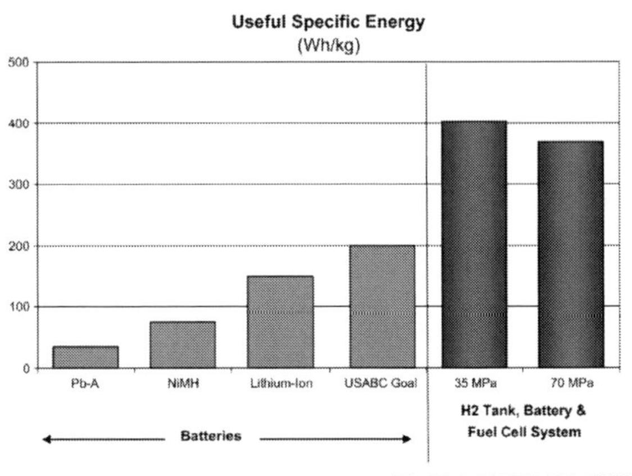

Figure 11. [32] Comparison of useful specific energy density of different batteries with two types of hydrogen fuel tanks.

The above graph is clearly explaining the compressed fuel cell, which can provide the energy upto five times more than the nickle-metal hydride and two times more than the Li-ion based batteries. In order to increase the range of the FCEV; very small amount of change in the size of fuel tank is done. While, in case of Electric Vehicles; the large size of batteries are added; which also increases the overall payload capacity and size of the motors of a vehicle. The FCEV tanks take more space than the gasoline vehicles based tanks. The 70 MPa tanks are preferred by the auto companies to increase the density of the stored energy. The Comparison of energy density is also shown in below Figure 12.

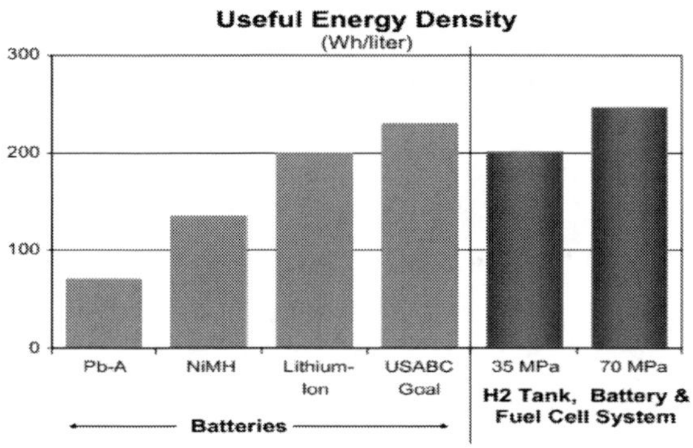

Figure 12. [32] Comparison of useful energy density of different batteries with two types of hydrogen.

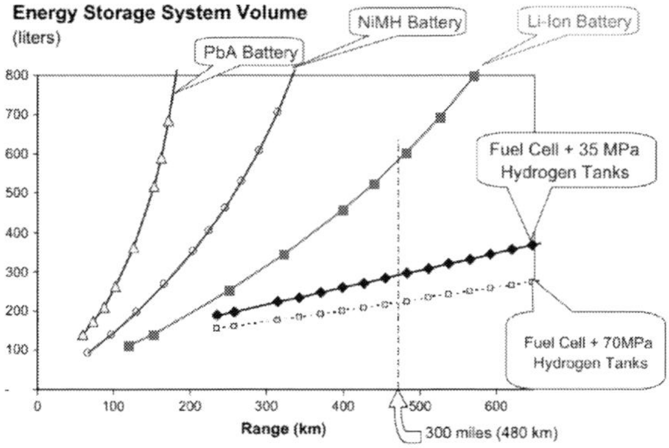

Figure 13. [32] Parameters that affects the range of a vehicles.

Total volume require for FCEV fuel cell with tank and different batteries are given in Figure 13. From the figure, it is clear that; for a long range(greater than 80km) vehicle the weight of a battery is very high than the weight of the fuel cell and tank combined together. For the same range despite having same energy, the Electric Vehicles are having more weight compare to the Fuel Cell Electric Vehicle.

Based on the Refueling Time

Refueling time is a major factor to differentiate the factors on the evaluation of a battery. Because, the FCEV need hydrogen to operate; and in this case only the work done is required to fill the hydrogens within a tank. While, the EV need to charge a battery which takes electrical power from an energy sources. This difference in refueling method gives a big advantages to the fuel cell based vehicles. It also takes the transfer hydrogen from one tank to another at a very high speed [33]. While, the battery is having a limit of power transfer capacity due to which it cannot be charged too fast.

The latest battery which is Li-ion battery cannot be charged above a certain limit due to safety reasons; Even if the batteries are made to charge rapidly there is one more barrier to be overcome, which is known as the \Local electric power supply limit." There are basically three types of power supply which are Level 1 (120V, 20A, 2KW), Level 2 (240V, 40A, 8KW) and fastest is Level 3 which can deliver upto 80-150KW of power. For a 320Km of a range, 82KWh battery is needed. At level 1 charging unit takes 42h to charge and at level 2 unit it takes 11h to charge the vehicle. At level 3 it may take vehicle to charge within a 36-84 mins. This concludes that refueling time for an EV is significantly higher than a fuel cell based conveyances [32].

Based on the Costs

The cost of a 100 watt power fuel-cell is 205USD/kW. While, the total expenses of control electronics, storage tank, battery management system, motors and wheels of a fuel cell propulsion systems costs about 29147USD [34]. For BEV, the cost of a Lithium-ion battery is estimated to be 250USD/kWh [35]; combining it with other components of the car that is electric drivetrain (motors, inverter, BMS etc.) costs about: 19951USD [36]. The Overall cost of a battery operated vehicle is less than FCEV, but looking at high range vehicle from figure 8, the cost of a BV (Battery Vehicles) increases significantly with respect to the range of a vehicles. Whereas, the cost of a FCEV does not increase with respect to the cost of a BEV.

Challenges of Hydrogen Fuel Cell

As increasing the number of prototypes, the fuel cell technology keeps proceeding at a brisk rate. More engagement is shown by Europe and USA from the last few decades to advance the fuel cell technology and disseminate its future growth in the transportation industry. However commercialization of fuel cell technology still lingers a challenge. Also cost of fuel cell, durability and reliability, infrastructure development and hydrogen storage are some of the biggest issues hindering the development of hydrogen powered vehicles.

Barriers in Commercialization of Fuel Cells

The three factors i.e., cost, quality and acceptance by the end-user are the key necessity to commercialize any product. The public and end-user would accept fuel cell technology as an eco-friendly alternative despite the quality and cost being comparable to conventional gasoline vehicles. This acceptance has led to large investments for subsidies of the fuel cells in recent years. High costs of hydrogen production, reliability and durability are the greatest hindrance to the fuel cell technology [37]. For the last 50 years, research and development from a large number of private and government organizations have been done. But, this kind of issues for hydrogen infrastructure development is still unsolved [38].

Cost of Fuel Cell

The cost of fuel cells can be divided as the material cost, labor cost, component cost and capital cost of manufacturing equipment. The mass of a material used in the component helps to determine the material cost [48]; which is dependent on the technological advancements, the cost of membrane, catalyst and bipolar plates; which keeps varying in nature. The 46 g of platinum is required for 50 kW of fuel cell; this kind of requirement is called as a platinum loading which acts as a key component of the hydrogen fuel cell. With development of fuel cell technology, analysts expect that the loading of platinum per car will reduce significantly; that reduces the cost of platinum loaded per car. As per the changes in demand, the possible impact of rising prices of platinum is a main barrier for the commercialization of fuel cell vehicles [39]. Compared to the conventional gasoline engines, the fuel cells are stationary under working conditions. So the cost of operation of fuel cell and its fabrication should be less. Developing cheaper materials and catalysts with high operating temperature can be one of the methods to reduce the costs of fuel cells [37].

Technical Barriers

Major technical barriers for a fuel cells are the - reliability and durability within a system; even though there are many successful pilot projects regarding the fuel cells; but no organization or company yet has proven a durable, reliable and cost-efficient fuel cell; which can compete and replace the traditional gasoline engine technology. Also, for acceptance from user end, reliability becomes the first important issue after durability as the fuel cell stack will avoid disintegration before the repair and maintenance within a specific period of time [40]. The accumulated degradation of catalyst and materials due to water and heating issues create the problems for low durability and reliability [37].Reliability can be defined as the ability of fuel cell to execute the specified condition under declared condition for a period of time; the reliability also leads to a catastrophic failure in a system. While, the durability can be defined as the ability of proton exchange membrane fuel cell to hold out against enduring change in performance over time; also the durability doesn't lead any kind of catastrophic failure within a system [41].

Due to the poor water management, corrosion and chemical reactions of cell components; the membranes, gas diffusion layers and electrodes of a system gets damaged [42]. The enhancement in the lifetime of a fuel cell without changing the materials and catalysts of a system is done by controlling the flow condition of a system. Owing to the - no moving parts in the fuel cell, it may lead the degradation of catalysts and materials in a system [43]. As per the above mentioned conditions, the proper water and thermal management are the main parameters for enhancing the durability and reliability of a system. Apart of it, the operational costs of fuel cell are much higher than the diesel price and hence the fuel cells need conceptual breakthroughs to overcome with this challenge [44].

Hydrogen Infrastructure

Managing the massive demand of hydrogen fuel cell based equipment's and its availability with respect to the needs and type of applications have always been a challenging task for mankind. With this regard, the installations of sundry kind of hydrogen based generating stations are incrementing drastically, which generates the sufficient amount of hydrogens with respect to demand of the loads. These loads could be anything that uses hydrogen energy as an input for performing any specific task. And among them, the Infrastructures of a hydrogen based fuel cells plays a very crucial role for performing and consummating the needs of sundry kind of load. The astute and efficient charging infrastructure will not only consummate the needs of various loads, but additionally truncates the overall cost function of the system.

As mentioned in M. Steen [44], the development of generating stations, transportations, protections, and distribution of hydrogen using the pipelines; plays an important role for developing the efficient and smart hydrogen infrastructure. The generating stations are used to generate the hydrogen for loads, while the transportation system supplies those generated hydrogen with respect to various distributed stations. Once the hydrogen reaches to the distribution centres, the loads are attached and filled with hydrogen to perform its specific (defined) tasks.

As mentioned in - Romero, S., et al. [45]; The hydrogen fuel has been transferred from the production house to the distribution center using three basic techniques which has been elucidated in below Figure 14 [46]. As per the figure, the generated hydrogen within a production house has been transferred in two ways. The first method is - by using a vehicle/transportation system; while the other method is used to transfer the hydrogen by using a various kind of pipelines. In this process, two kind of special truck has been used for transforming the hydrogen from generating (production house) station to the receiving (distribution) station. For compressed hydrogen, special kind of Compressed Gas (CG) based Truck is used; while in case of liquid hydrogen, the Liquid Hydrogen (LH) based Truck is used for transporting the hydrogens.

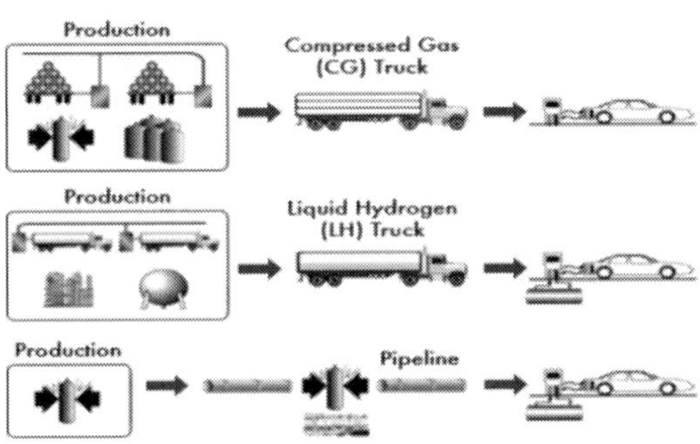

Figure 14. [46] Hydrogen distribution pathways.

Sometimes, the compressed gaseous hydrogen at a pressure of (40-80) bar is supplied using a pipeline based transportation system. This kind of transformation not only help to deliver the large volume of hydrogen; but also reduces the overall cost function of the system [57]. Due to the increasing transport distance, liquified hydrogen is used for its higher density, which is helpful for storing the larger mass in small container; which reduces the overall cost function required in transportation [47].

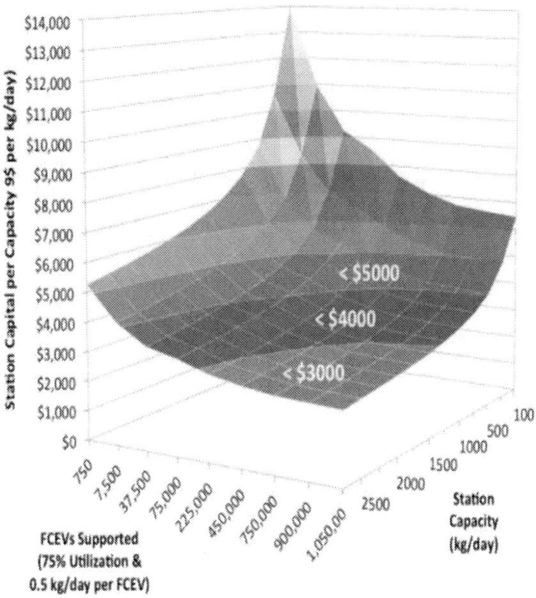

Figure 15. [48] Capital costs required for hydrogen stations.

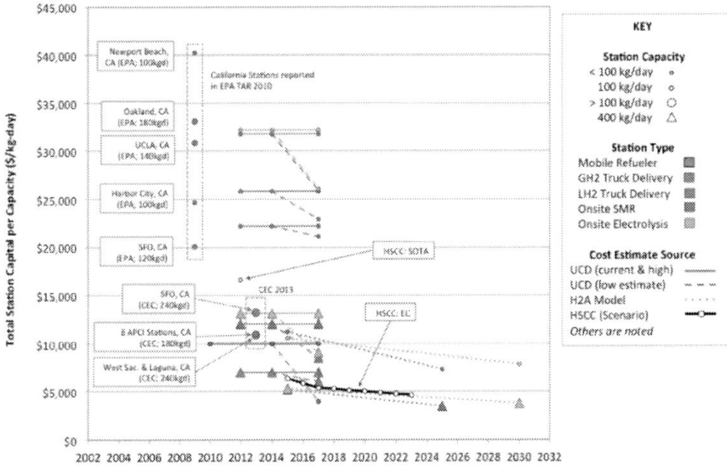

Figure 16. [48] Shorter term capital costs per capacity of hydrogen stations over time (in years).

As per the above figures; the capital cost required for hydrogen based stations is described with the help of Figure 15; the shorter and longer term capital.

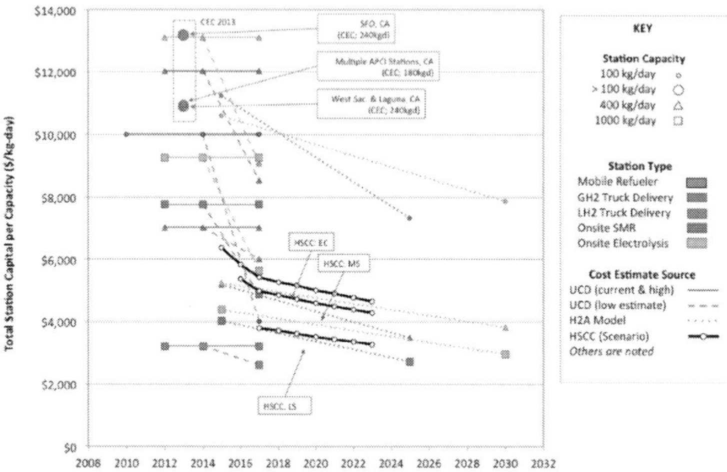

Figure 17. [48] Longer term capital costs per capacity of hydrogen stations over time (in years).

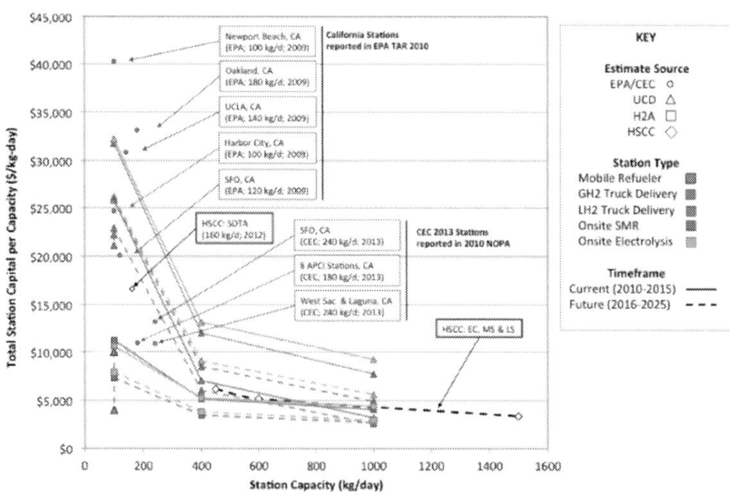

Figure 18. [48] Capital cost per capacity as a function of station capacity in (kg/day).

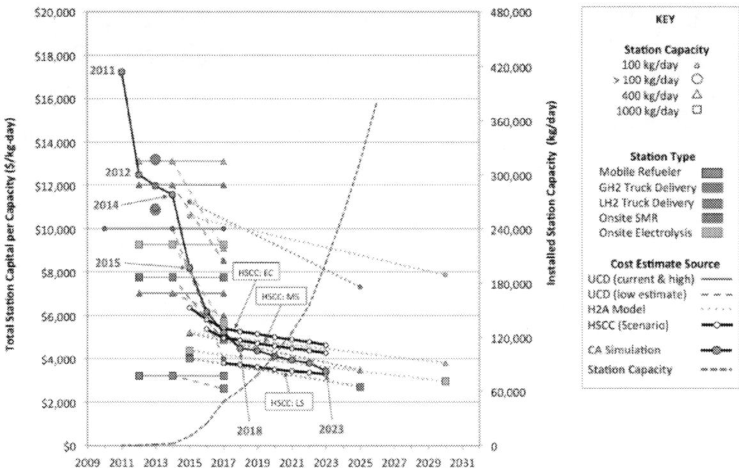

Figure 19. [48] Reduction in station capital costs per capacity for the national scenario of a California early market.

FUTURE WORKS

Storing the huge amount of hydrogen fuel is always been a challenging task. So scientists across the world discovered that formic acid which is a good candidate of storing the hydrogen fuel; but required a lot of heat and energy to produce hydrogen from it. So researchers developed a new simple method by using the palladium-based nanomaterial which is quite efficient than old conventional based storing technologies [49]; Hence, more research is required in field of storage contrivances for a hydrogen fuel cell. It can also be stored in many ways to reduce the chances of inflation, such as gas at normal pressure or at high pressure, in the form of liquid hydrogen or as a solid hydride. Liquifying hydrogen is very difficult so there is need of research on this [50]. It is also cheap to transport hydrogen via a pipeline. There is no need of whole new construction of pipelines specially for hydrogen. Also, it is possible to reuse pipelines of natural gas by utilizing the retrofitting method [51]. Although the cost of a green hydrogen and blue hydrogen is decrementing drastically; but for making it more financially friendly researchers are working on the conception on the production of hydrogen in vehicles. The cost of transportation will additionally get minimized by utilizing this kind of technologies [51]. There is an alternative method to compose hydrogen; as the extracting hydrogen from water is more power consuming. Therefore, hydrogen can be composed by reacting the carbon monoxide with water, but this is less eco-friendly [52]. By compressing platinum, it has 90% ability to reduce oxygen in water which may further increase the overall efficiency of hydrogen fuel cells [53]. Additionally reducing the utilizations of precious metals as a catalyst and increasing the use of non-precious metals will reduce the cost of the fuel cells along with incrementing the production and usage of fuel cells [54].

CONCLUSION

This review explicates the working of the fuel cells, its types, comparative study, technical barriers and future aspects along with the avail of recent research work from different researchers and surveys in detail. Hydrogen shows a promising future to propel the transportation industries estimated by the past studies predicated on the hydrogen powered conveyances. The costs of hydrogen fuel, its transportation charges, reduction in prices of platinum used as a catalyst and search for its alternates are the few points which required a deep research in future to commercialize the hydrogen fuel cell in an efficient manner. The mechanical durability is a consequential factor to enhance the performance of a fuel cell conveyances; while the advancements and adoption from the user end is affected by the costing of the fuel cell vehicles. The FCEV needs 46g of platinum for every 50KW of fuel cell. But the researchers have prognosticated that the prices of platinum may hike as per the production of proton exchange membrane increases. So there is a need of research to use an alternative catalyst in the PEM fuel cell; as it also helps to reduce the cost of the fuel cell which is used in the vehicles. The FCEV's having three times more specific energy than the electric vehicle and the calorific value of hydrogen is 141.80MJ/kg; compare to the calorific value of petrol which is near about 45.8MJ/kg. The biggest advantage of FCEV's are that they emit water as their output; after using the hydrogen for propelling the vehicle which eventually leads to reduce the global warming effect and produces an eco-friendly fuel. In this paper, the comparative study of electric vehicle and hydrogen fuel cell vehicle has been discussed in detail. Which shows the energy output of compressed fuel is five times more than the electric vehicle powered by nickel-metal hydride battery; while two times more than the EV powered by Li-ion battery packs. The topic of hydrogen infrastructure in this paper estimates the different ways of hydrogen transportation with utilization of felicitous techniques and vehicles to transfer hydrogen from production house to distribution center safely. The overall hydrogen powered vehicles show a promising future in the transportation industry as a clean and range extender fuel and reducing the dependency on fossil fuels. Also the future research topics to be focused in the field of FCEV's are enhancing the cold-start performance, life of fuel cell, enhancing the membrane materials and fuel storing devices and possible ways to increase the range of fuel cell conveyances have additionally been highlighted in detail.

Conflict of Interest

This research did not receive any specific grant from funding agencies in the public, commercial, or not for profit sectors.

References

[1] Acar, Canan, Ibrahim Dincer, and Greg. 2016. "Clean hydrogen and power from impure water." *Journal of Power Sources* 331: 189-197.

[2] Stougie, Lydia. Natalìa, Giustozzi. Hedzer van der, Kooi, and Anna, Stoppato. 2018. "Environmental, economic and exergetic sustainability assessment of power generation from fossil and renewable energy sources." *International Journal of Energy Research* 42(9): 2916-2926.

[3] Zhang, Xin-Rong. 2017. *Natural future of energy utilization*. 757-760.

[4] Dincer, Ibrahim, and Canan, Acar. 2017. "Smart energy systems for a sustainable future." *Applied Energy* 194: 225-235.

[5] Cipriani, Giovanni, et al. 2014. "Perspective on hydrogen energy carrier and its automotive applications." *International Journal of Hydrogen Energy* 39(16): 8482-8494.

[6] Salvi, B. L. and Subramanian, K. A. 2015. "Sustainable development of road transportation sector using hydrogen energy system." *Renewable and Sustainable Energy Reviews* 51: 1132-1155.

[7] Ehret, Oliver, and Klaus, Bonhoff. 2015. "Hydrogen as a fuel and energy storage: Success factors for the German Energiewende." *International Journal of Hydrogen Energy* 40 (15): 5526-5533.

[8] Acar, Canan. Sayantan, Ghosh. Ibrahim, Dincer, and Calin, Zamfirescu. 2015. "Evaluation of a new continuous type hybrid photo-electrochemical system." *International Journal of Hydrogen Energy* 40(34): 11112-11124.

[9] Alrafea, Kamal. Michael, Fowler. Ali, Elkamel, and Amir, Hajimiragha. 2016. "Integration of renewable energy sources into combined cycle power plants through electrolysis generated hydrogen in a new designed energy hub." *International journal of hydrogen energy* 41(38): 16718-16728.

[10] Chan, Ching Chue. 2007. "The state of the art of electric, hybrid, and fuel cell vehicles." *Proceedings of the IEEE* 95(4): 704-718.

[11] Wee, Jung-Ho. 2007."Applications of proton exchange membrane fuel cell systems. "*Renewable and sustainable energy reviews* 11(8)(2007): 1720-1738.

[12] Mekhilef, Saad. Rahman, Saidur, and Azadeh, Safari. 2012."Comparative study of different fuel cell technologies." *Renewable and Sustainable Energy Reviews* 16(1): 981-989.

[13] Hydrogen Technology, 2020. *Hydrogenics.* Last modified April 20. https://www.hydrogenics.com/technology-resources/hydrogen-technology/fuel-cells/.

[14] Hacker, Viktor. Alexander, Schenk. Simon, Fraser, and Karl, Kordesch. 2015. "Cells and Stacks." In *Reference Module in Chemistry, Molecular Sciences and Chemical Engineering.* Elsevier BV.

[15] Breeze, Paul. 2018. "Hydrogen energy storage." In: *Power System Energy Storage Technologies.* In Elsevier: 69-77, https://doi.org/10.1016/b978-0-12-812902-9. 00008-0.

[16] Strath, 2020. *Hydrogen Economics*. Last modified April 25. http://www.esru.strath.ac.uk/EandE/Web_sites/0203/hydrogen_economy/Types%20of%20Fuel%20Cells.htm .

[17] American History, 2020. *Fuel Cells*. Last modified April 26. https://americanhistory.si.edu/fuelcells/phos/pafcmain.htm

[18] Fuller, T. F. and Gallagher, K. G. 2008. "Phosphoric acid fuel cells." In *Materials for fuel cells*: 209-247.

[19] Ma, L. D. Ingham, B. Pourkashanian, M. and Carcadea, E. 2005. *Review of the computational fluid dynamics modeling of fuel cells.* 246-257.

[20] Hydrogen Economy, 2020. *Economic Growth.* Last modified April 26. http://www.esru.strath.ac.uk/EandE/Web_sites/0203/hydrogen_economy/Types%20of%20Fuel%20Cells.htm.

[21] Scott, K. Taama, W. M. and Argyropoulos, P. 1999. "Engineering aspects of the direct methanol fuel cell system." *Journal of Power Sources* 79(1):43-59.

[22] Methanol, 2020. *Dornsif.* Last modified April 28. https://dornsife.usc.edu/labs/narayan/methanol-fuel-cells/.

[23] Rozendal, R. A. T. H, J. A. Sleutels, H. V. Hamelers, M. and Buisman, C. J. N. 2008. "Effect of the type of ion exchange membrane on performance, ion transport, and pH in biocatalyzed electrolysis of wastewater." *Water Science and Technology* 57(11): 1757-1762.

[24] Fu, Kun Kelvin. Yunhui, Gong. Jiaqi Dai, Amy Gong. Xiaogang Han, Yonggang Yao. Chengwei Wang. 2016. "Flexible, solid-state, ion-conducting membrane with 3D garnet nanofiber networks for lithium batteries." *Proceedings of the National Academy of Sciences* 113(26): 7094-7099.

[25] Pnas, 2020. *Medium Org*. Last modified April 29. https://www.pnas.org/content/pnas/113/6/7094/F1.medium.gif.

[26] Hallinan Jr, Daniel T. and Nitash P. Balsara. 2013. "Polymer electrolytes." *Annual review of materials research* 43 (2013): 503-525.

[27] Grigoriev, S. A. et al. 2011. "High-pressure PEM water electrolysis and corresponding safety issues." *International journal of hydrogen energy* 36(3): 2721-2728.

[28] Peighambardoust, S. Jamai, Soosan Rowshanzamir, and Mehdi, Amjadi. 2010. "Review of the proton exchange membranes for fuel cell applications." *International journal of hydrogen energy* 35(17): 9349-9384.

[29] Niya, Seyed Mohammad Rezaei, and Mina Hoorfar. 2014. "Process modeling of the ohmic loss in proton exchange membrane fuel cells." *Electrochimica Acta* 120: 193-203.

[30] Berrueta, Alberto. Idoia San, Martín. Pablo, Sanchis, and Alfredo, Ursúa. 2019. "Lithium-ion batteries as distributed energy storage systems for microgrids." In *Distributed Energy Resources in Microgrids*: 143-183.

[31] File: images, 2020. *Image*. Last Modified April 29. https://pubs.rsc.org/image/article/2017/ee/c6ee03499c/c6ee03499c-f1 hi-res.gif.

[32] Thomas, C. E. 2009. "Fuel cell and battery electric vehicles compared." *International Journal of Hydrogen Energy* 34(15): 6005-6020.

[33] Granovskii, Mikhail. Ibrahim, Dincer, and Marc A. Rosen. 2006. "Economic and environmental comparison of conventional, hybrid, electric and hydrogen fuel cell vehicles." *Journal of Power Sources* 159(2): 1186-1193.

[34] Cuenca, R. M. Gaines, L. L. and Vyas, A. D. 2000. *Evaluation of electric vehicle production and operating costs*. No. ANL/ESD-41. Argonne National Lab., IL (US).

[35] Gaines, Linda, and Roy, Cuenca. 2000. *Costs of lithium-ion batteries for vehicles*. No. ANL/ESD-42. Argonne National Lab., IL (US).

[36] Eaves, Stephen, and James, Eaves. 2004. "A cost comparison of fuel-cell and battery electric vehicles." *Journal of Power Sources* 130(1-2): 208-212.

[37] Wang, Junye. 2015. "Barriers of scaling-up fuel cells: cost, durability and reliability." *Energy* 80: 509-521.

[38] Behling, Noriko. 2012. "Solving the fuel cell dilemma." *Fuel Cells Bulletin* 2012(11): 12-14.

[39] Sun, Yongling. Mark, Delucchi, and Joan, Ogden. 2011. "The impact of widespread deployment of fuel cell vehicles on platinum demand and price." *International Journal of Hydrogen Energy* 36(17): 11116-11127.

[40] Wang, Junye. 2015. "Theory and practice of flow field designs for fuel cell scaling-up: a critical review." *Applied Energy* 157: 640-663.

[41] Wu, Jinfeng. Xiao Zi Yuan, Jonathan J. Martin, Haijiang. Wang, Jiujun, Zhang. Jun Shen, Shaohong. Wu, Walter Merida. 2008. "A review of PEM fuel cell durability: Degradation mechanisms and mitigation strategies." *Journal of Power Sources* 184(1): 104-119.

[42] Schmittinger, Wolfgang, and Ardalan, Vahidi. "A review of the main parameters influencing long-term performance and durability of PEM fuel cells." *Journal of power Sources* 180, no. 1 (2008): 1-14.

[43] Sun, Xinwei. Kaiqi, Xu. Christian, Fleischer. Xin, Liu. Mathieu, Grandcolas. Ragnar Strandbakke, Tor S. Bjorheim, Truls Norby, and Athanasios, Chatzitakis. 2018. "Earth-abundant electrocatalysts in proton exchange membrane electrolyzers." *Catalysts* 8(12): 657.

[44] Steen, M. 2016. "Building a hydrogen infrastructure in the EU." In *Compendium of Hydrogen Energy*, pp. 267-292. Woodhead Publishing.

[45] Campíñez, Romero, et al. 2018. "A hydrogen refuelling stations infrastructure deployment for cities supported on fuel cell taxi roll-out." *Energy* 148: 1018-1031.

[46] Larminie, James, et al. 2003. *Fuel cell systems explained.* Vol. 2. Chichester, UK: J. Wiley.

[47] Vickers, Neil J. 2017. "Animal communication: when I'm calling you, will you answer too?." *Current Biology* 27(14): R713-R715.

[48] Cost, Hydrogen Infrastructure. *Presentation Overview*. (2005).

[49] Xiong, Huiyuan, et al. 2019. "An energy matching method for battery electric vehicle and hydrogen fuel cell vehicle based on source energy consumption rate." *International Journal of Hydrogen Energy* 44(56): 29733-29742.

[50] Felseghi, Raluca Andreea, et al. 2019. "Hydrogen fuel cell technology for the sustainable future of stationary applications." *Energies* 12(23) 4593.

[51] Ahluwalia, Rajesh K. Xinglian, Wang. Aymeric, Rousseau, and R, Kumar. 2004. "Fuel economy of hydrogen fuel cell vehicles." *Journal of Power Sources* 130(1-2): 192-201.

[52] Granovskii, Mikhail. Ibrahim, Dincer, and Marc A, Rosen. 2006. "Economic and environmental comparison of conventional, hybrid, electric and hydrogen fuel cell vehicles." *Journal of Power Sources* 159(2): 1186-1193.

[53] Frenette, Greg, and Daniel, Forthoffer. 2009. "Economic & commercial viability of hydrogen fuel cell vehicles from an automotive manufacturer perspective." *International Journal of Hydrogen Energy* 34(9): 3578-3588.

[54] Ogden, Joan M. Margaret, M. Steinbugler, and Thomas G. Kreutz. 1999. "A comparison of hydrogen, methanol and gasoline as fuels for fuel cell vehicles: implications for vehicle design and infrastructure development." *Journal of Power Sources* 79(2): 143-168.

In: Research Challenges in Science … ISBN: 978-1-68507-008-3
Editors: S. Kannadhasan et al.

Chapter 14

MEDISEARCH: A FRAMEWORK FOR KEYWORD SEARCH ON MEDICAL TERMS RELATED TO AIR POLLUTION

Bhimavarapu Usharani*

Department of Computer Science and Engineering
Koneru Lakshmaiah Education Foundation
Vaddeswaram, India

ABSTRACT

Air pollution is one of the major environmental issue of the developed and developing countries throughout the world. Around 90 percent people in the world are breathing the contaminated air. The universal ecological pollution is assessed as a world-wide public health concern with numerous issues. Amongst all the clinical search queries on the Internet, air pollution associated health issue data is one of the utmost wide spread issues that originates much of the society's attention, specifically on the globe. There is the need of similarity-based technique to create word interpretations and to assess clinical terms. This paper discusses work on keyword search a kind of "medical keyword search" which

* Corresponding Author's E-mail: ushareddy.vja@gmail.com. Orchid ID: https://orcid.org/0000-0001-5050-0415.

can help users to find medical terms on the internet. MediSearch combines Google web API with a hierarchy visualization technique. It permits the user to perform keyword search on medical terms using medical related files and to outwardly examine the documents to check their pertinence and get the natural remedies for air pollution related health issues.

Keywords: air pollution, ozone, indexer, keyword search, query search, medical keywords etc.

INTRODUCTION

Now a days, internet is one of the essential sources for the people, to get data regarding health, illnesses, and remedial treatment. A report by the World Health Organisation (WHO) in 2018[1] disclosed that inhaling in the polluted air put down around 600,000 infants every year below the age of 15 years. A description by the Centre for Science and Environment (CSE) in 2017 [2] discovered that thirty percent of all premature mortalities in India are triggered anticipated to contaminated air. Energy Policy Institute of university of Chicago (EPIC) [3] specified that India is the earth's second most contaminated country. The Centre for Science and Environment (CSE) in 2019 [2] discovered that chronic obstructive pulmonary disease (COPD) due to air contamination at forty nine percent is accountable for close to half of the deaths, traced by lung cancer deaths at thirty three percent, heart diseases and diabetes at twenty two percent each and stroke at fifteen percent. A survey, issued in the journal Environmental Health Perspectives, [4] observed that short-term interaction to in house air pollution was correlated together with aggravations of psychiatric syndromes. Air pollution affects hormonal activity and shows its impact on menstrual cycle irregularity.

Even strong and healthy people also can face ailments from contaminated air comprising respiratory irritation or breathing complications during workout or open-air activities. Those most vulnerable to acute ailments from air pollution are:

- Individuals with artery disease, heart failure or heart disease.
- People with lung illnesses allergies or chronic obstructive pulmonary disease.
- Open-air workers.
- Senior people and the aged.
- Kids below 14 years.
- Sportspersons those who exercise enthusiastically outdoors.

An one-keyword search query is assumed as a conventional keywords of size one, Q = {m1, m2, …, mn}, and look for interrelated rows that have the assumed key-words, where a row includes a keyword if a content characteristic comprises the key-word.

This paper describes that the MediSearch framework is an air pollution based medical keyword search queries on a database evolving and exploitation a full-content search engine intended to enable information retrieval (IR), functions analogous to Google but is furnished with distinct features for supervision of tasks distinct to repossessing data from medical content. To counterpart present applications, developed MediSearch, which is a medical text search engine for looking biomedical perceptions that are potentially related to a query. The unique advantage of MediSearch is that it sends real-time replies while being able to receive flexible queries.

MediSearch accepts a query from the user as the input. The system then repossesses all the pages that match the query from MediSearch using term and notion guides. The concepts limited in the pages are then calculated and ranked rendering to their importance to the query.

The main contributions of this research are:

1. MediSearch, is a robust and easy-to-use open-source keyword search infrastructure for air pollution health issue relevant terminology.
2. MediSearch uses a procedure to diminish the duplication that is formed by the traditional execution using the pragmatic page methodology.

3. MediSearch integrates a medical spell checker to aware manipulators about possibly misspelled words in their exploration queries.
4. A study of the matters associated with indexing and ranking structured information.

The remainder of the article is arranged as follows: Section 2 examines about the air pollution and their health impacts. Section 3 discusses about the already existing techniques in the medical search for keyword query frameworks. Section 4 explains the MediSearch method architecture. Section 5 discusses the experimental results and their performance. Finally, the conclusion and are given in the Conclusion section.

AIR POLLUTION AND THEIR HEALTH IMPACTS

According to the World Health Organization (WHO)[1], the contaminants that pose the high risk to an individual's health are:

1. Nitrogen Dioxide (NO_2)
2. Sulphur Dioxide (SO_2)
3. Ozone|(O_3)
4. Particulate Matter ($PM_{2.5}$ or PM_{10})
5. Carbon Monoxide (CO)

Nitrogen Oxide (NO2)

This is the exacerbation of the breathing procedure as it pierces cavernous in the lung, bringing lung ailments, bronchospasm, coughing, dyspnoea, breathless, and unfluctuating pulmonary edema when breathe in at excessive levels. It is mentioned that the extreme stages of nitrogen dioxide can be responsible for lasting diseases. Extensive breathing of NO2 can harm the feeling of odour and taste [5, 6, 7].

Sulphur Dioxide (SO2)

Sulphur dioxide is an injurious gas, discharged largely as of fossil coal depletion or industrial movements [8]. It disturbs individuals, animal, and plant life. Vulnerable human beings as these with historic people, lung disease, and children, who pose a complicated chance of damage. The foremost well-being troubles associated with sulphur dioxide discharges in industrial zones are breathing annoyance, secretion creation, bronchitis, and bronchospasm, as it is a bodily irritant and pierces subterranean into the lung changed into bisulphite and touching on with sensory receptors. Furthermore, pores and skin soreness, mucous membranes, damage to the eyes, and worsening of pre-existing cardiac disorder have been spotted [7].

Ozone

Ozone acceptance typically happens by way of inhalation. Ozone disturbs the tear ducts and the higher layers of the pores and skin [9]. A learn about of quick exposure of extraordinary quantities of ozone exhibited ailments formation in the higher pores and skin however additionally discount in nutritional vitamins C and E. It is probably that ozone ranges are now not intrusive with the pores and skin barrier position and related to instantaneous to pores and skin ailment [10]. Due to the quick water-solubility of ozone, panted ozone can pierce extremely into the lungs [11]. Ozone awareness associated to arise in the everyday quantity of losses, in the variety of breathing fatalities, and in the range of heart casualties.

Particulate Matter

Particulate matter comprises minute liquid or strong precipitations that can be breathed and reason risky fitness outcomes [12]. Particles <10 μm in diameter (PM10) after breath can assault the lungs and even unfold the bloodstream. Long-term publicity to PM for years used to be originated to be associated to coronary heart illnesses and early mortality.

Carbon Monoxide (CO)

The indications of destroying due for gasping carbon monoxide incorporate headache, nausea, vomiting, weakness, and, finally, loss of perception. The negative impact of carbon monoxide to haemoglobin is far larger than that of oxygen. In vein, hazardous intoxication may additionally perchance manifest in individuals bare to extreme concentrations of carbon monoxide for an extensive phase of time. Due to the loss of oxygen due to the fact of the aggressive binding of carbon monoxide, ischemia, hypoxia and coronary heart sickness are detected. [13, 14]. Pollution habitually originates from unsafe pollution and can also lead to in sickness eye infection [15], retinopathy [16,17].

LITERATURE SURVEY

These days cyberspace is appropriate for one of the generally essential resource to get medical and physical condition data for the public [18]. The Pew Internet & American Life Project stated [19] that around 80% of web users search for medical or health-related data from end to end the Internet. It has been bring into being that the utmost frequent health topics look for on the Internet were a precise illness or medical setback, diet, nutrition, specific medical healing, vitamins, and workout or vigour [19], as well as treatment drugs [20]. Users choose to search keyword-based queries in their native languages, while often queries in these human languages cannot be accurately understood by system.

Keyword queries stated by means of keywords are the slightest sensitive one, since it is signified as a set of terms without any obvious relation between them. In information retrieval systems the frequency of explore results depends on the individual ability to signify their data needs in a query. Ordinary language representation of query affords added data than the keyword-based method since a linguistic analysis can be accomplished to excavation syntactic information [21]. Keyword search queries like BANKS [22], DBXplorer [23] and Discover [24], the handler affords as input a set of keywords and the computer returns an answer a set of rows [25].

MEDISEARCH

Considered a relational database representation as a coordinated graph GS (V, E), called a schema graph, {R1, R2, ⸺, Rn} where V addresses the arrangement of connection schemas {V1,V2,... Vn.} and E implies the arrangement of edges concerning connection patterns.. Relational database can be viewed as an information base diagram GD (V, E) on the chart GS V portrays a bunch of columns, and E addresses a bunch of edges between rows. An n-keyword is designated as a set of key phrases of dimension n, Q = {k1, k2, •••,kn}, explorations organised rows that comprise the targeted key phrases, the place a row comprises a key-word if a textual content attribute of the row encompasses the keyword.

The furthermost common way of using MediSearch framework is to enter the keywords into an unpretentious text entry box. Search terms may comprise of single words or multi-word expressions, wild cards, and other operators. Stop words are conserved in the page catalogues because several are valid abbreviations of medical perceptions.

Medical jargon comprises numerous difficult-to-spell phrases which can be inspiring even to veteran clinicians. It is additionally now not rare for misspellings to make their way into approved affected person records. These can be confrontations incorrectly spelled such as “choallara” as an alternative of “cholera” or phrases that had been spelled accurately however have been nonetheless improper. The comprehensive planning of the Medisearch is represented in Figure 1:

To discourse these topics, Medisearch includes a clinical spell checker to conscious customers about perchance misspelled confrontations in their exploration queries. In accumulation, Medisearch proposals a preference for handlers to contain possibly misspelled techniques of the search phrases in the search. The execution of these points was once primarily based on phonetic comparable and order evaluation algorithms supplied in open-source spell look at APIs. Moreover, spell mistakes, Medisearch additionally examines for frequent common-sense blunders originate in user-submitted search queries, e.g., identical key phrases seeming on each presence and removal lists. The interrelation between the tables is confirmed in Figure 2:

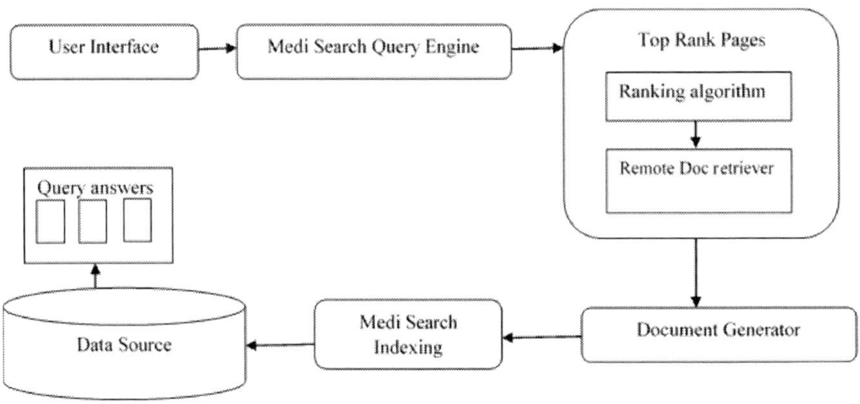

Figure 1. MediSearch Architecture.

Did	Dname	Sid
d1	asthma	s1,s3

sid	symptoms
S1	coughing, especially at night, when laughing, or during exercise
S2	tightness inthe chest
S3	shortness of breath
S4	difficulty talking
S5	anxiousness or panic
S6	fatigue

Mid	Medicine
M1	Fluticasone
M2	Budesonide
M3	Mometasone
M4	Beclomethasone
M5	Ciclesonide

tid	did	mid
T1	D1	M1
T2	D2	M3

Figure 2. An example database.

Row Generator

A table is a customary row linked over primary or foreign keys and provisions all the data unearthed from the correlated rows. The hyperlinks signify the semantic associations between units of rows. Therefore, the row generator shops the family member's records between the rows and comprehends key phrases that can be used in a query.

Document Generator

This generator is accountable for producing the files by means of excruciating the rows, perceiving, and reducing superfluous facts to produce a set of special and condensed documents. The document generator algorithm is given under in algorithm 1:

Algorithm 1: Document Generator Algorithm

Input: An m-keywordsearch query Q = {k1, k2, …, km}

Output: Set of redundancy reduced documents

1. Get all the nodes.
2. Get all the neighbours of the nodes in the graph Gs
3. Select conventional nodescollectedfromdesignated nodes from the Row generator.
4. Collect documents that keep the text stored in the rows that are dynamically assembled rows.
5. Eliminate the duplicate information from the document, which creates the unique document.

Indexer

To prepossess data MediSearch makes use of an inverted index, known as Document Generation Inverted Index (DGII), which is a better model of a conventional inverted index. The Indexer provisions the DGII index. The index accesses in DGII are the key phrases restrained in the document generator. Each key-word factors to an incline of record generator the place the keyword is confined, and each document generator incorporates a related rating that specifies the weight that this key-word has in the file generator. The document generator in every listing remains organized in descending order taking into model their scores. An easy way to rating and exuberant the Document Generator is through the utilization of some deviations of the TF IDF metric. Fagin Procedure [35] and Threshold Procedure [42] are two of the most significantly used methods for responding top-k queries over an organized list. Implemented the Fagin algorithm to classify the top-k responses on topmost of the DGII Index.

RESULTS

All the algorithms had been applied in the programming language Java and the research has done on a laptop with an Intel Core CPU, four GB RAM on Windows 10.The further most close-fitting end result in assessing a realistic informatics device such as Medisearch is possibly may human beings practise the device in their day-to-day work.

$$\text{Recall}=\frac{\text{No of Corresponding items retrieved}}{\text{No:of items in the collection}} \tag{1}$$

$$\text{Precision}=\frac{\text{No of Corresponding items retrieved}}{\text{Total no:of items retrieved}} \tag{2}$$

MediSearch is in contrast with three diverse approaches (i) a technique depend on Meta-Map (MM+)[4] (ii) phrase abstraction and Support Vector Machine classification (TT-SVM) and (iii) a closing manner the use of the supervised learning with CRF (BIO-CRF), which is then united with the scheme MM+ by way of the use of the results of MM+ as semantic topographies (BIO-CRF-H). The contrast consequences are given in Table 1:

Medical keyword associated exploration on semi-structured and structured datasets, constructed on a document generator method, suggestions extreme effectively and unique effects when coping with average measurement datasets, this method presented restrictions outstanding specially to an incompetent loading utilization essential via facts laying-off, a massive index shape and an exact time-consuming directory creation. This approach reduces redundancy and evades record duplicity. A document retrieval was once made to equivalence the variety of remote documents produced, the time to generate the index and the measurement of the index. Designated one twenty queries with numerous numbers of keywords. In this experiment, used 20 queries with two to six keywords affecting a whole of one hundred queries. These outcomes are confirmed in Figure 3. To arrange query, diverse arbitrary quantity of key phrases used to be special for each attribute in the tables. Top-k clinical keyword search effectively on dataset the usage of 5 key phrases is proven in Figure 4. Top-10 search precision and recall are proven in Figure 5 and Figure 6.

Table 1. Comparing the Medisearch results with the existing work

Project	**Category**	**Precison**	**Recall**	**F-score**
MM+	Disease	60.84	53.04	56.67
	Symptoms	51.99	61.93	56.53
	Remedies	56.67	28.48	37.91
TT-SVM	Disease	48.25	43.16	45.56
	Symptoms	42.45	50.86	46.28
	Remedies	57.37	35.76	44.06
BIO-CRF-H	Disease	82.05	73.14	77.45
	Symptoms	83.18	73.33	78.12
	Remedies	87.50	68.69	77.07
Medisearch	Disease	90.5	85.7	89.9
	Symptoms	91.3	85.8	89.1
	Remedies	93.8	86.2	89.5

Table 2. Indexing time (ms) for the two approaches

	MM+	TT-SVM	BIO-CRF-H	Medisearch
Indexing time	100	100	90	80

Table 3. Size (in MB) for the two approaches

	MM+	TT-SVM	BIO-CRF-H	Medisearch
Index size	7345	7041	7405	5389

This approach reduces redundancy and evades report duplicity. Whit the usage of Document Generator, Medisearch can generate and install massive scale indexes except humiliating effectively and accuracy. Indexing time and dimension of more than a few present works are given in Table two and Table 3.

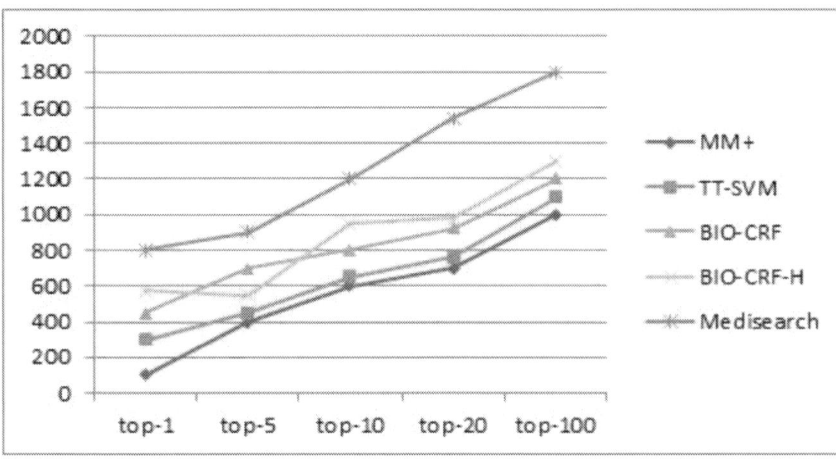

Figure 3. Medical Keyword Efficiency for top-100 results, using two to six keywords on dataset.

The metric precision is employed, which produces the ratio of the variety of associated solutions amongst the first okay question solutions with the upper most scores.

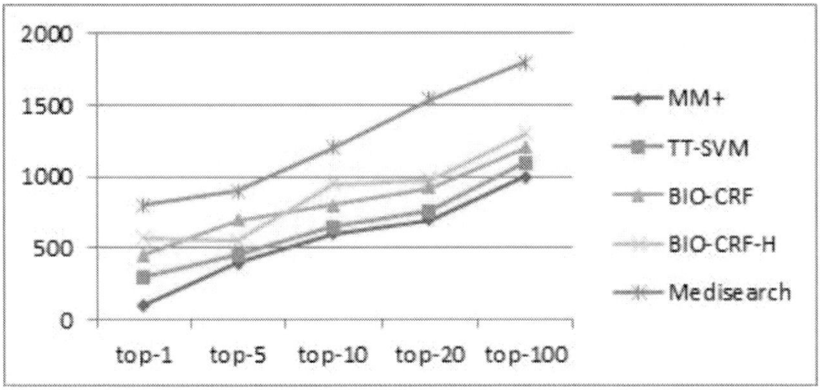

Figure 4. Top-k medical keyword efficiency on dataset using 5 keywords.

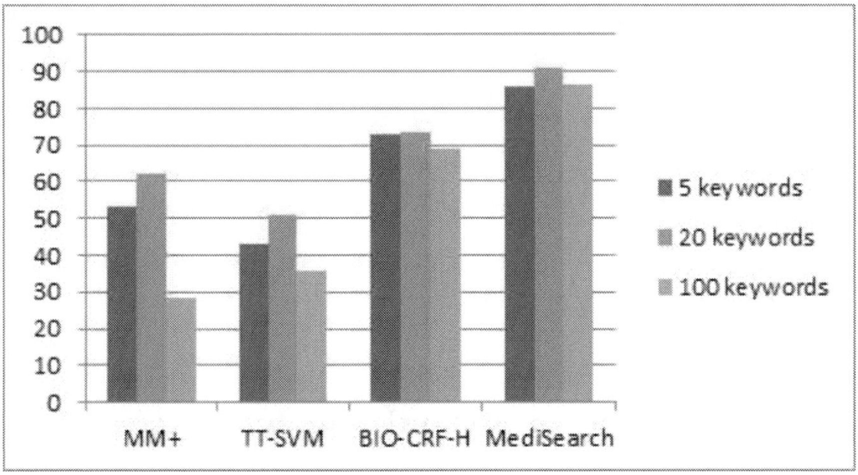

Figure 5. Top-10 search precision on dataset.

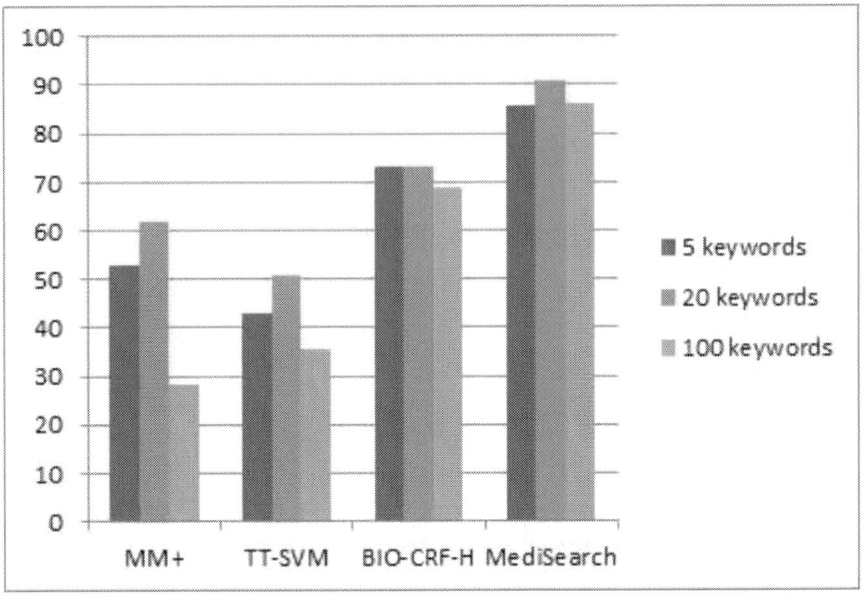

Figure 6. Top-10 search Recall on dataset using keywords.

CONCLUSION

Human happenings have an opposing consequence on the ecosystem by means of contaminating the water we take, the air we inhale, and the ground in which flowers raise. Air pollution consists of trivial particles that can be regular or false and candidate stress humans each out-of-doors and at home. The results of air pollution on individual's fitness can assortment from minor respiratory problems to extreme cardiac issues, such as coronary heart disorder and stroke. Healthcare professionals and scholars remain to search for environment friendly and high-quality informatics equipment to provision their necessities for repossessing statistics from description of affected person records. Medisearch search engines, working like Google API however furnished with elements to adapt the distinct features of scientific text, providing a realistic answer for assembly such needs.

References

[1] https://www.who.int/news-room/air-pollution, Online accessed on 02 February 2021.

[2] https://www.cseindia.org, online accessed on 30 January 2021).

[3] https://aqli.epic.uchicago.edu.

[4] https://ehp.niehs.nih.gov.

[5] Richmont-Bryant J, Owen RC, Graham S, Snyder M, McDow S, Oakes M, 2017. Estimation of on-road NO2 concentrations, NO2/NOX ratios, and related roadway gradients from near-road monitoring data. *Air Qual. Atm. Health.*

[6] Hesterberg TW, Bunn WB, McClellan RO, Hamade AK, Long CM, Valberg PA. 2009, Critical review of the human data on short-term nitrogen dioxide (NO_2) exposures: evidence for NO2 no-effect levels. *Crit. Rev. Toxicol.* (2009) 39:743–81.

[7] Chen T-M, Gokhale J, Shofer S, Kuschner WG. 2007 Outdoor air pollution: nitrogen dioxide, sulfur dioxide, and carbon monoxide health effects. *Am. J. Med. Sci.* () 333:249–56.

[8] US EPA. *Table of Historical SO_2 NAAQS, Sulfur US EPA*. Available online at: https://www3.epa.gov/ttn/naaqs/standards/so2/s_so2_ history.html.

[9] WHO *Health Risks of Ozone From Long-Range Transboundary Air Pollution*. Available online at: http://www.euro.who.int/data/assets/ pdf_file/0005/78647/E91843.pdf (accessed JANUARY 15, 2021).

[10] Thiele JJ, Traber MG, Tsang K, Cross CE, Packer L. 2017, *In vivo* exposure to ozone depletes vitamins C and E and induces lipid peroxidation in epidermal layers of murine skin. *Free Radic. Biol. Med.*23:365–391.

[11] Hatch GE, Slade R, Harris LP, McDonnell WF, Devlin RB, Koren HS, 2014. Ozone dose and effect in humans and rats. A comparison using oxygen- 18 labeling and bronchoalveolar lavage. *Am. J. Respir. Crit. Care Med.* 150:676–683.

[12] Cheung K, Daher N, Kam W, Shafer MM, Ning Z, Schauer JJ, I. 2011Spatial and temporal variation of chemical composition and mass closure of ambient coarse particulate matter (PM10–2.5) in the Los Angeles area. *Atmos. Environ.* () 45:2651–62.

[13] Kuan Ken Lee, Nicholas Spath, Mark R. Miller, Nicholas L. Mills, Anoop S.V. Shah 2020, Short-term exposure to carbon monoxide and myocardial infarction: *A systematic review and meta-analysis Environment International,Volume 143,* October 2020.

[14] Chen, RJ., Lee, YH., Chen, T H. et al.,2021,Carbon monoxide-triggered health effects: the important role of the inflammasome and its possible crosstalk with autophagy and exosomes. *Arch. Toxicol.* ().springer.

[15] Weisskopf MG, Kioumourtzoglou M-A, Roberts AL. 2015, Air pollution and autism spectrum disorders: causal or confounded? *Curr. Environ. Health Rep.* () 2:430–9. 10.1007/s40572-015-0073-9.

[16] Mo Z, Fu Q, Lyu D, Zhang L, Qin Z, Tang Q, 2019. Impacts of air pollution on dry eye disease among residents in Hangzhou, China: a case-crossover study. *Environ. Pollut.* () 246:183–9. 10.1016/j.envpol.2018.11.109.

[17] Klopfer J. 2019 Effects of environmental air pollution on the eye. *J. Am. Optom. Assoc.* () 60:773–8.

[18] Taylor H, Leitman R. eHealth traffic critically dependent on search engines and portals. *Healthcare News 2001*;13:1-3.

[19] Fox S. Health information online. Washington, DC: *Pew Internet & American Life Project;* 2005 May 17. URL: http://www.pewinternet.org/Reports/2005/Health-Information-Online/1-Summary-of-Findings.aspx, online accessed on 04 January 2021.

[20] Law MR, Mintzes B, Morgan SG. 2020 The sources and popularity of online drug information: an analysis of top search engine results and web page views. *Ann. Pharmacother.* Mar.;45(3):350-356.

[21] Aronson, A. R. (2021). Effective mapping of biomedical text to the UMLS metathesaurus: *The MetaMap program* (Vol. 8, pp. 17–21).

[22] G. Bhalotia, A. Hulgeri, C. Nakhe, S. Chakrabarti, S. Sudarshan, 2012Keyword searching and browsing in databases using banks, in: *Proceedings of the 18th International Conference on Data Engineering,* San Jose, CA, USA, February 26–March 1, pp. 431–440.

[23] Agrawal S., S. Chaudhuri, G. Das, 2002 Dbxplorer: a system for keyword-based search over relational databases, in: *Proceedings of the 18th International Conference on Data Engineering (ICDE),* San Jose, CA USA, February 26–March 1, pp. 1163–1166.

[24] Hristidis V., Y. Papakonstantinou, 2002 Discover: Keyword search in relational databases, in: *Proceedings of the 28th International Conference on Very Large Databases*, August 20–23, Hong Kong, China, pp. 670–681.

[25] S.Kannadhasan, G.Karthikeyan and V.Sethupathi, A Graph Theory Based Energy Efficient Clustering Techniques in Wireless Sensor Networks. *Information and Communication Technologies Organized by Noorul Islam University (ICT 2013) Nagercoil on 11-12 April 2013, Published for Conference Proceedings by IEEE Explore Digital Library* 978-1-4673-5758-6/13 @2013 IEEE.

In: Research Challenges in Science ... ISBN: 978-1-68507-008-3
Editors: S. Kannadhasan et al.

Chapter 15

AUTOMATIC HIBISCUS LEAF DISEASE DETECTION AND CLASSIFICATION USING UNSUPERVISED LEARNING TECHNIQUES

Bhimavarapu Usharani*

Department of Computer Science and Engineering,
Koneru Lakshmaiah Education Foundation,
Vaddeswaram, AP, India

ABSTRACT

Hibiscus is a fantastic herb and in Ayurveda it is one of the most renowned herbs that have extraordinary healing properties. Hibiscus is rich in vitamin C, flavonoids, amino acids, mucilage fibre, moisture content and antioxidants. Hibiscus can help with weight loss, cancer treatment, bacterial infections, fever, high blood pressure, lower body temperature, treat heart and nerve diseases. Automatic leaf disease detection is an essential task. Image Processing is one of the popular techniques for the plant leaf disease detection and categorization. In this paper, the diseased leaf is identified by Concurrent K-Means Clustering

* Corresponding Author's E-mail: ushareddy.vja@gmail.com. Orchid ID: https://orcid.org/0000-0001-5050-0415.

algorithm and then features were extracted. Finally, Reweighted KNN Linear Classification algorithms have been used to detect the diseased leaves categories.

Keywords: concurrent K-Means, reweighted KNN linear, segmentation, hibiscus etc.

1. INTRODUCTION

Hibiscus is widely known for its beautiful flowers which lack aroma. Hibiscus belongs to the Malyaceae group, and its scientific name is Hibiscus rosa-sinensis. A popular name for hibiscus in ayurvedic is Japa. This hibiscus plant does not have any fruit. Hibiscus flowers are in various shades like red, white, yellow, pink. The bark, leaves and flowers are known to have medical properties.

Hibiscus is a famous herb found all over India in the temperature climates and its other types are found all over the world. Hibiscus reduces high blood pressure, using for cancer treatment and as a weight loss aid, along with other uses. Research has revealed a variety of health benefits linked to drinking hibiscus tea, lowers blood pressure, fight bacteria and even aid weight loss. Hibiscus extracts boosted the number of antioxidant enzymes and lessened the damaging impacts of available radical by up to 92% [1]. The article [2] associated the outcomes of drinking hibiscus contrasted with black tea on fat stages. The authors in [3] presented that hibiscus followed in a body mass index (BMI), weight of a person, overweight of a person. The authors of [4, 5] originate that drinking hibiscus decoction dropped blood pressure in people at risk of high blood pressure and those with slightly high blood pressure. The authors in [6] discussed that this herbal is used to treat lice [6]. The results of [7] proved that hibiscus leaf extract prevented human prostate cancer. Hibiscus leaves extract inhibits stomach cancer cell by up to 52%. Hibiscus extracts are used in many medical applications like cramping, gas, diarrhea [9], to treat bacterial infections [10].

Rest of this chapter is categorized as following sections: Section II of this article briefly discusses about existing work for the plant disease detection. Section III presents the proposed model i.e., Automatic

Hibiscus Leaf Disease Detection and Classification Using Unsupervised Learning Techniques. Section IV presents experimental and performance results and finally section V presents the concluding remarks.

2. LITERATURE SURVEY

Leaf images are one of the most important resources for the recognition and categorisation of plant groups and their diseases. WanMohdFadzil et al. [11] conversed a bug identification process for orchid plant shrubberies. The proposed procedure used the combination of several approaches of edge segmentation methods, morphological and filtering procedures cast-off for classifying given images into two bug class as black leaf spot and solar scorch.

Rong Zhou et al. [12], explained method to identity leaflet patch in sugar beet. The procedure implements cross methods for corresponding and support vector machinel (SVM). This technique uses R for segmentation, filtering and support vector machine (SVM) classifier for categorization.

Dheeb et al. [13] proposed a technique to spot the leaf disorder of the plant that exists on flyer and bamboo. It uses a combination of K-Mean's technique and deep learning for categorization of the input leaf and the segmented images, it gives a maximum precision of 93%.

Smith et al. [14] projected a procedure for identification of plant leaf maladies and the exactness of the proposed technique is verified by comparing the taken input leaf images with those segmented leaf images. Di et al. [15] proposed a technique to detect erosion on leaf and its progression of bug and uses the concept of centroid. Youwen Tian and Lin Zhang [16] proposed a technique for cucumber downy mildew bug by using the hyperspectral imaging. Muralil et al. [17], proposed a procedure to identify the filthy plant sprig area using the K-Means and extract the region of interest.

3. Procedure for Hibiscus Leaf Disease Detection and Classification

The architecture of proposed model is depicted in Figure 1 which shows the broad process of projected hibiscus leaf disease detection. The steps in the proposed system are:

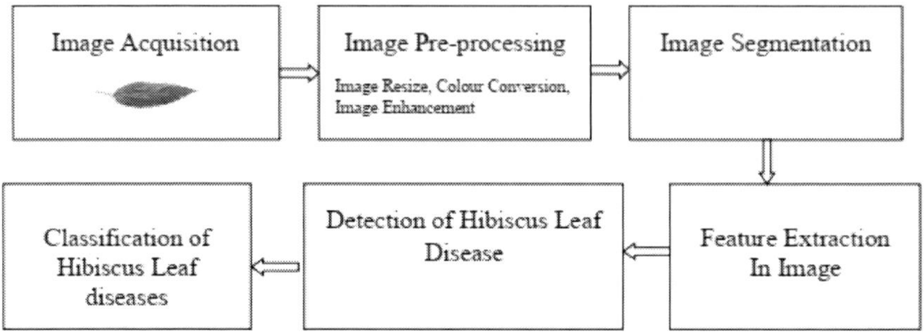

Figure 1. Procedure for hibiscus leaf disorder detection.

3.1. Image Acquisition

The hibiscus leaf images are got by using the mobile camera. The dataset comprises 150 healthy and diseased hibiscus leaf images. Dataset consists of two set of classes: the first class is the healthy hibiscus leaf images, and the second leaf images set is the unhealthy hibiscus leaf image set. The healthy and the infected leaves samples are depicted in Figure 2 and Figure 3.

Figure 2. Healthy Hibiscus Leaves.

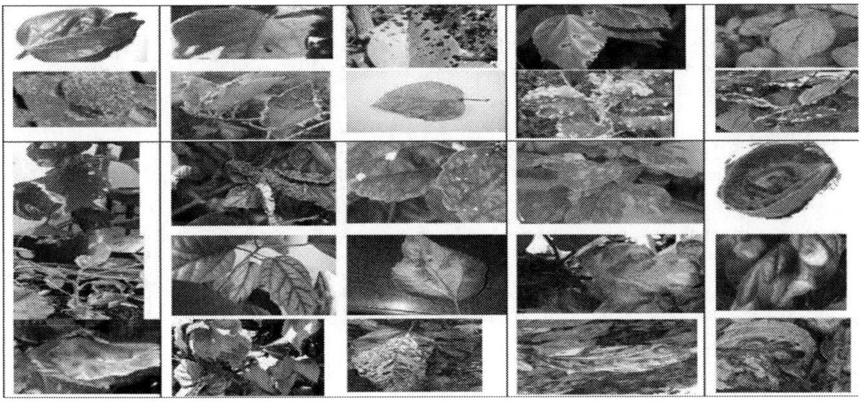

Figure 3. Diseased and Infected Hibiscus Leaves.

3.2. Image Preprocessing

The preprocessing is applied on the hibiscus leaf image to make it suitable for further processing. The first technique in the preprocessing is the resize the given input hibiscus leaf image by converting to sizes of 512X512 images. Next converted Red Green Blue i.e., RGB format to Hue Saturation Value i.e., HSV format. Later apply the image enhancement technique. The resultant image is depicted in Figure 4:

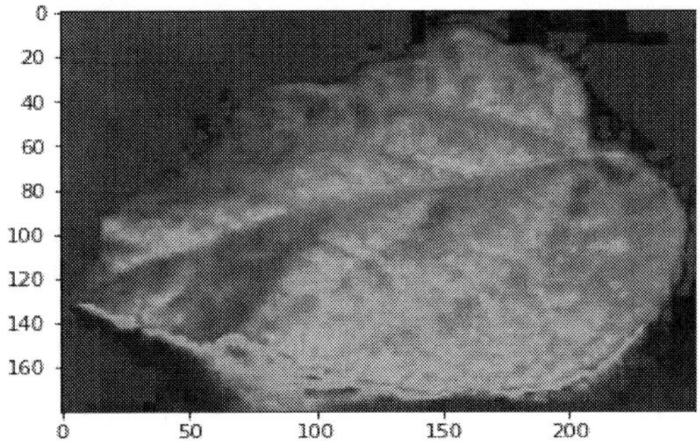

Figure 4. Pre-processing

3.3. Image Segmentation

Segmentation divides the image into several regions. It analyses the hibiscus leaf image data set to extract the information for further processing. In this article the segmentation uses the Concurrent K-Means clustering technique. The proposed algorithm is discussed in Algorithm 1.

Algorithm 1: Concurrent K- Means Clustering Algorithm

1. Divide the dataset into different sets and each set is clustered concurrently.
2. Initialize each cluster centroid.
3. Until the stopping criteria met.
4. Recalculate the new cluster centroid by using the normalized euclidean distance.
5. Merge all the cluster centroids.
6. Map the points to final cluster.

3.4. Feature Extraction

Feature extraction extracts the features that are present in the hibiscus leaf images. The extracted features are categorized as shape, color, and texture. The leaf disease varies their shapes into different several shaped of the image due to existing diseases. The color feature differentiates the disease from each other.

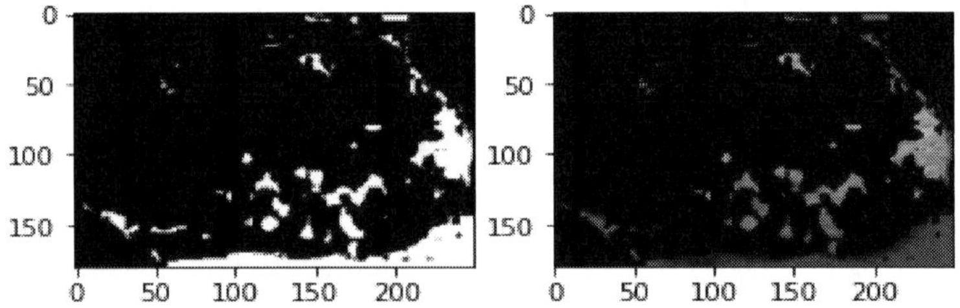

Figure 5. Extracting the features for the hibiscus leaf.

3.5. Classification

Classification categorizes every input hibiscus leaf image into a specific class which improves the leaf disease detection accuracy. The proposed Reweighted KN-Linear classifier reduces the incorrect classification of healthy or infected hibiscus leaf images. The algorithm for the reweighted KNN linear classifier is given in Algorithm 2.

Algorithm 2: Reweighted KNN Linear Classification

Input: Set of Hibiscus Leaf Images

Output: Improve detection accuracy

1. For each leaf image $L_{i.}$
2. Extract the features.
3. Construct n weak learners D_i
4. Classify the leaf as normal or infected.
5. Combine all weak learners result $y=\sum(D_i)$
6. Calculate training error $T_{e.}$
7. Reweight the weak n- learners.
8. Find best weak learner with minimum training error.
9. Obtain classification results.

4. Results and Discussion

The hibiscus leaf image of plant is captured by using the mobile in the Vaddeswaram and Vijayawada locations. The taken input hibiscus leaf images comprises of 150 images of healthy, infected hibiscus leaf leaves. The acquired leaf images are converted into the useable format. All the experiments are implemented using the python.

To classify hibiscus leaf disease, a set of characteristics for classification and detection of the diverse hibiscus infected plants are considered. For this, leaf spot characteristics are extricated from image using the suitable image processing methods and these characteristics stipulate serious data about the pictorial illustration. The characteristics correspond to features of spots are length, ratio of the principal axes which describes the ratio of the principal axis's length.

Orientation is demarcated as an approach among the major axis of the spot and the horizontal axis. The equivdiameter is the diameter of the spot. The eccentricity is the ratio of the distance between the foci and the major axis length of the ellipse eccentricity ratio. Solidity tells about the compactness of the spot. The extent metric gives the details of the proportion of the pixels in the infected area. The hydraulic radius gives the about by isolating the spot area with the spot perimeter. The complexity metric gives the perimeter of the infected leaf spot area. The Euler number gives the invariant property of the infected area of the leaf. The above discussed metrics are given for the hibiscus diseases and given in Table 1 and Table 2.

Eccentricity value must be in between 0 and 1, the spot whose eccentricity ratio is 0 is a circle. The extent metric value must be in concerning as zero and one, when the proportion of the spot has the value 1 then its shape is impeccably rectangle. By using the feature extraction, the outcome for numerous characteristics of the hibiscus plant images is extracted and summarized in Table 1 and Table 2.

Figure 6 shows the results of finding the affected area of hibiscus leaves with different diseases. Identified 14 various hibiscus leaf diseases affected.

Table 1. Features of hibiscus leaf diseases

Feature	Dieback	Wilt	Leaf fungus	Black Spot	Psedomonas Cichorii	Myrothecium Roridum
Area	76.03	65.34	78.11	69.78	65.34	77.54
Major axis	18.70	13.45	11.34	12.45	11.45	11.78
Minor axis	9.23	8.01	7.98	6.34	8.34	7.56
Orientation	34.89	43.56	23.56	34.12	23.11	29.48
Equivdiameter	10.78	11.89	10.34	9.89	10.67	10.67
Eccentricity	0.98	0.89	0.91	0.93	0.91	0.95
Solidity	0.95	0.91	0.94	0.96	0.95	0.91
Extent	0.78	0.87	0.71	0.69	0.75	0.77
Hydraulic radius	1.98	1.56	1.27	1.34	1.67	1.45
complexity	15.67	11.23	13.67	12.45	11.56	14.38
Euler Number	1.00	1.00	1.00	1.00	1.00	1.00
Moment of inertia	14.56	13.45	12.25	14.27	13.67	11.71
Ratio of Principle axes	1.78	1.25	1.67	1.83	1.41	1.49

Table 2. Features of hibiscus leaf diseases

Feature	White Powdery Mildew	Yellow Leaves	Phytophthora	Bacterial Blights	Leaf Spot	White cottony spots
Area	78.34	69.26	65.49	77.54	76.03	65.34
Major axis	11.45	11.78	13.45	11.34	12.45	11.49
Minor axis	7.98	6.34	9.23	8.01	7.34	8.01
Orientation	34.12	23.11	34.89	43.56	41.23	39.34
Equivdiameter	11.89	10.34	10.67	10.67	9.56	10.45
Eccentricity	0.91	0.93	0.91	0.95	0.91	0.94
Solidity	0.94	0.96	0.95	0.93	0.91	0.95
Extent	0.87	0.71	0.71	0.69	0.75	0.77
Hydraulic radius	1.34	1.67	1.98	1.56	1.27	1.34
complexity	11.23	13.67	15.67	11.23	13.67	12.45
Euler Number	1.00	1.00	1.00	1.00	1.00	1.00
Moment of inertia	14.27	13.67	13.45	12.25	14.27	13.67
Ratio of Principle axes	1.67	1.83	1.31	1.59	1.88	1.15

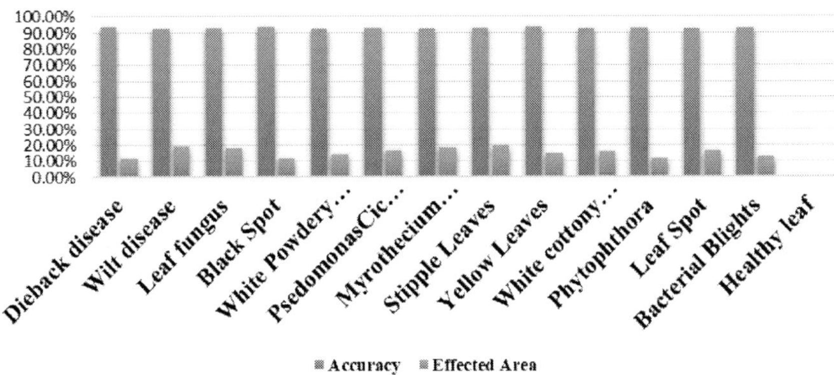

Figure 6. Disease names, affected area, and their accuracy.

Conclusion

It is very important to diagnose the infected plant leaves automatically. To segment the leaf, concurrent K-Means clustering algorithm has been proposed. By this the parallel execution of cluster centroids has taken place and reduces the complexity and computational cost. The feature extraction has done using the geometrical features. Then finally reweighted KNN linear classification algorithm has been proposed to classify the category of the diseased hibiscus leaves. The given system gives the 92.15% accuracy for the hibiscus leaf disease.

References

[1] Taofeek O Ajiboye, Nasir A Salawu, Musa T Yakubu, Adenike T Oladiji, Musbau A Akanji, and Joseph I Okogun. 2011, Antioxidant and drug detoxification potentials of hibiscus sabdariffa anthocyanin extract. *Drug and chemical toxicology*, 34(2):109–115.

[2] Abbas Mohagheghi, Shirin Maghsoud, Patricia Khashayar, and Mohammad Ghazi-Khansari., 2011, The effect of hibiscus sabdariffa on lipid profile, creatinine, and serum electrolytes: a randomized clinical trial. *International Scholarly Research Notices* 2011.

[3] Hong-Chou Chang, Chiung-Huei Peng, Da-Ming Yeh, Erl-Shyh Kao, and Chau-Jong Wang. 2014 Hibiscus sabdariffa extract inhibits obesity and fat accumulation and improves liver steatosis in humans. *Food & function*, 5(4):734–739.

[4] Diane L McKay, Edward Saltzman, Chung-Yen Chen, and Jeffrey B Blumberg. *2018 Hibiscus sabdariffa l. tea (tisane) lowers blood pressure in prehypertensive and mildly hypertensive adults.*

[5] Corina Serban, Amirhossein Sahebkar, Sorin Ursoniu, Florina Andrica, and Maciej Banach. 2015 Effect of sour tea (*Hibiscus sabdariffa* l.) on arterial hypertension: a systematic review and meta-analysis of randomized controlled trials. *Journal of hypertension*, 33(6):1119–1127.

[6] Zeinab M El-Basheir and MA Fouad. 2012A preliminary pilot survey on head lice, pediculosis in sharkia governorate and treatment of lice with natural plant extracts. *Journal of theEgyptian Society of Parasitology*, 32(3):725– 736.

[7] Hui-Hsuan Lin, Hui-Pei Huang, Chi-Chou Huang, Jing-Hsien Chen, and Chau-Jong Wang. *2015 Hibiscus polyphenol-rich*

extract induces apoptosis in human gastric carcinoma cells via p53 phosphorylation and p38 mapk/fasl cascade pathway. Molecular Carcinogenesis: Published in cooperation with the University of Texas MD Anderson Cancer Center, 43(2):86–99.

[8] Chun-Tang Chiu, Jing-Hsien Chen, Fen-Pi Chou, and Hui-Hsuan Lin. 2005Hibiscus sabdariffa leaf extract inhibits human prostate cancer cell invasion via down-regulation of akt/nf-kb/mmp-9 pathway. *Nutrients*, 7(7):5065– 5087.

[9] EunKyung Jung, YoungJun Kim, and Nami Joo. 2015 Physicochemical properties and antimicrobial activity of roselle (hibiscus sabdariffa l.). *Journal of the Science of Food and Agriculture*, 93(15):3769–3776.

[10] Emad Mohamed Abdallah. 2016 Antibacterial efficiency of the sudanese roselle (*Hibiscus sabdariffa* l.), a famous beverage from sudanese folk medicine. *Journal of intercultural ethnopharmacology,* 5(2): 186.

[11] Wan Mohd Fadzil, shah Rizam, Jailani R., Nooritawati M.T, 2014,Orchid leaf disease detection using Border egmentation technique, *IEEE conference on systems, Procescontrol* (ICSPC), 1, 168-179.

[12] Rong Zhou, Shunichi kaneko, Fumio Tianaka, Miyuki Kayamori, Motoshingeshimizu, 2013,Early detection and continuous quantization of plant dieseaseuisng template matching and support vector machine algorithms, *IEEE International symposium on computing and networking*.

[13] Ai-Bashish,D.M Braik and S.Ban-Ahmad, 2011 Detection and classification of leaf diseases using K-Means based segmentation and neural networks-based classification, *Inform Technol*,10,267-275,

[14] Amargo, A, C., J.S Smith, 2019 *An Image processing-based algorithm to automatically identify plant disease visual symptoms*, Elsevier, 17(1),9-21.

[15] Di Cui,Qin zhang,Mizan LI, Glen L.Hartman, Youfu Zhao, 2012 *Image Processing methods for quantitatively detecting soybean rust from multispectral images*, Elsevier, 11(5),743-750.

[16] Youwen Tian, Linzhang, *2012 Study on the methods of detecting cucumber downy mildew using hypersectral imaging technology,"* Elsevier,11(5),743-750.

[17] Murali Krishnan, Dr. M. G sumithra, 2013, A Novel algorithm for detecting bacterial leaf scorch (BLS) of shade trees using image processing, *IEEE 11 Malaysia international conferenceon communications.*

In: Research Challenges in Science ... ISBN: 978-1-68507-008-3
Editors: S. Kannadhasan et al.

Chapter 16

Recent Trends in Solar Energy Generation, Challenges and Opportunities

***Guruprasath Rengarajan*[1,*] *and Alagar Karthik*[2,†]**

[1]Department of Electronics and Communication Engineering, Gnanamani College of Technology,Namakkal, Tamlinadu, India,
[2]Department of Electrical and Electronics Engineering, KPR Institute of Engineering and Technology, Coimbatore, Tamilnadu, India

Abstract

Solar energy innovations have seen phenomenal development in recent years. This strong involvement in these innovations has been encouraged and maintained by the realization of technical advances, rising public knowledge of environmental problems, the economic situation, and the number of policy instruments. Although the cost of solar-generated electricity is still costly and the power from renewable energies, like solar, is still low energy, there has been no large-scale production of renewable resources

*Corresponding Author's E-mail: jcetguru@gmail.com.
†Corresponding Author's E-mail: karthick.power@gmail.com.

and generation providers are still least involved in purchasing power from renewable sources. This paper gives an analysis of the technological, economic and policy implications of the production of solar energy.

Keywords: solar energy, solar energy project and solar photovoltaic

1. INTRODUCTION

Solar energy functions in terms of resource potential, present capabilities, historical patterns, and future solar energy development prospects.The scientific, economic and structural obstacles to solar energy technology growth and use. The analyses current fiscal and regulatory policy instruments to promote the production of solar energy, showing the progress of these policy instruments in achieving their objectives. Under several situations, India electricity market was reformed. Energy stability is critical for achieving national goals for long-term development and improving people's quality of life. For much of the 21st century, world dependency on crude oil may persist, but continued reliance on crude oil is burdened with inherent market uncertainty related to limited global instability. Moreover, environmental warming is still a big problem, mainly induced by greenhouse gas pollution from fossil fuel electricity production systems. There is, thus, a need to establish alternative energies, such as non-conventional sources, given the two issues listed above. This solar energy, one of the non-conventional sources of power, and various methods of transforming it to power.

2. SOLAR ENERGY

Solar energy, such as heat and power, may be transformed into other sources of energy. People are using the sun's energy for a lot of things today.In remote locations which are not connected to the electrical grid, PV systems are often used. Watches, calculators, and lighted road signs are often used to fuel them. Photovoltaic is the simple conversion of light at the atomic level into electricity. Some materials display a property known as the photoelectric effect that causes light photons to be absorbed and electrons released. An electric current results when these free electrons are captured, which can be used as electricity. Wide parabolic collectors are used to absorb solar fuel, which is usually used to heat the fluid that is contained in the storage tank. This heat energy is delivered with feedback to the boiler, which is converted into high-pressure steam energy. This steam energy is sent to the high-speed turbine hall, contributing to the turbine's prime-mover rotation. Because the turbine is attached internally to the alternator, the incoming mechanical energy from the turbine is translated by the alternator into electrical energy. In the condenser, steam is concentrated, and reuse, the feed water returns to the boiler. The heat of the condenser's cooling water may be used for some other purposes.

3. Solar Energy Project

Nevada Solar One is a modern solar thermal plant located near Boulder City, Nevada, with a 64 MW generating capacity. Parabolic troughs are used by Nevada Solar One as thermal solar concentraters, liquid heating tubes which serve as solar receivers. These solar receivers are custom coated tubes made of glass and steel, and the newly designed power plant uses some 19,300 of these four-meter-long tubes. In the Mojave Desert, there are many solar power projects that provide power to the power grid. The name assigned to nine solar power plants in the Mojave Desert, installed in the 1980s, is Solar Energy Generation Systems (SEGS). Such plants have a total capacity of 354 megawatts (MW), rendering them the world's largest solar power facility. There are still proposals in the Mojave Desert to construct several large solar plants.

The Indian power system is primarily focused on fossil fuels, relying on large indigenous coal deposits for around three-fifths of the country's power generation capacity. But the Indian government has taken many measures in recent decades to restrict the usage of energy dependent on fossil fuels while encouraging renewable energy production. Solar energy is the most plentiful resource of green energy possible, and yet its theoretically available capacity in most regions of the world is well beyond the actual overall availability of primary energy. Solar energy solutions are, as such, a crucial instrument for reducing global carbon pollution.

The broad variety of technologies accessible today is divided into passive and active technologies to harvest the energy of the light. Active technologies shaping the material of this paper are widely classified into photovoltaic and solar thermal technologies, where solar thermal can be further categorised into electrical and non-electrical solar thermal applications. Over the past decade, the demand for many solar energy technologies has seen drastic growth, particularly the growth of the demand for grid-connected PV systems and solar hot water systems. India is currently the world's fifth largest electricity generating region, with a capacity of 243 GWs currently built, of which 69.5 percent is from thermal, 16.5 percent from hydro, 2 percent from nuclear, and around 12 percent from renewable energy sources. While the Indian power sector has experienced a seven-fold increase in its installed capacity from 30,000 MW in 1981 to over 24,3028 MW by March 2014, but there is still a huge generation and demand gap in India, more generation plants need to be established that are preferable to renewable sources by government as well as various private participation.

The installed capacity of solar energy projects in India currently stands at approximately 3000 MW. By 2020, India plans to generate 20 GW of solar power. Although the cost of electricity from certain renewable energy technologies appears to be large relative to traditional energy technologies, the cost pattern of solar energy technologies has shown rapid declines in the recent past and the potential for major declines in the immediate future. In addition to expense, a variety of hurdles have been identified that seem to hinder the rapid growth of such technologies.

4. Solar Photovoltaic

Solar Photovoltaic (PV) is a way of producing electrical power by utilizing semiconductors that exhibit the photovoltaic effect to transform solar radiation into direct current electricity. Solar panels consisting of a series of solar cells comprising a photovoltaic film are used in photovoltaic power production. It is a system that transforms solar energy directly into a photovoltaic effect into electricity. F.C. first recognised photoelectric effects in 1839 from Becquerel. In this phenomenon, during the removal of energy from radiation, electrons are released from matter. By coating Selenium with an incredibly thin layer of gold in 1883, the first solar cell was made.

Bell Laboratories noticed in 1958 that Silicon (Si) doped with such impurities was very light-sensitive. This discovery culminated in the development of the first functional solar cell generated from materials that emit electrons when subjected to EM radiation with a sunlight conversion efficiency of ~6 percent. Monocrystalline silicon, polycrystalline silicon, amorphous silicon, cadmium telluride, and copper indium gallium selenide/sulphide are mainstream products presently used for photovoltaic.

The development of solar cells and photovoltaic arrays has progressed considerably in recent years because of the increased demand for renewable energy sources. The amount of power available from a solar cell relies on the solar energy being transformed into direct current electricity (DC) when light reaches the solar panels. From the panels, the direct current flows and is transformed into alternating current (AC) used by local electric utilities.

Finally, the power passes into transformers, where the voltage is enhanced for distribution to the transmission lines such that power can be delivered to households and companies through municipal electric utilities. A related set of solar panels is a photovoltaic array (or solar array). The power one module will create is seldom adequate to fulfill a home or company's requirements, so the modules are joined together to form an array. Using an inverter, most PV arrays transform the DC power produced by the modules into alternating current that can power lamps, engines, and other loads. Through thousands of monitoring mirrors (called heliostats) in around a two square mile area, power towers (also known as central tower power plants or heliostat power plants) absorb and concentrate the thermal energy of the light.

The middle of the heliostat area is where a tower sits. On a receiver that sits on top of the tower, the heliostats target intense sunlight. The concentrated sunlight heats molten salt up to over 1,000°F (538°C) inside the recipient. The hot molten salt then flows to a thermal storage tank where it is deposited, retaining a thermal quality of 98 percent, and finally pumped to a steam generator. A new approach to build compact arrays of small silicon solar cells may generate devices that do not suffer from these trade-offs. These microcell arrays are as effective as traditional solar panels and could be easier to build since they use far less silicon.Solar power is infirm power, and by building suitable storage facilities, measures are needed to render it firm power. Through announcing reasonable policy benefits, renewable power may even be made a feasible source of electricity. Re-powering must be part of every plan to scale up solar power capacity, since it is necessary to allow maximum use of high solar radiation sites that remain unused due to fewer government and investment company initiative and to update or substitute the old panels with new, wide and larger, more powerful ones.

1. The conversion performance of solar panels/plates is improved by the cleaning of these panels by newer equipment. To force dust and debris away from the surface of these devices, the system utilizes automatic dry-sweep.
2. Research and development is being carried out across the world to procure more and more electricity from different devices and technology. Japan created a fabric under this idea that is regarded as a solar cell fabric that can absorb the energy from sunlight when you are travelling by wearing this fabric made of cotton. This fabric is constructed of thin solar wafer cells woven in an elegant fashion. The created electricity would be able to charge mobile devices and other portable electronic gadgets. The thread used would be thicker and improve the fabric cloth's lifespan or toughness. The same concept can be embedded/used in the blind builders, and when sun rayeson these, some form of curtain will also generate electricity. In

cooperation with solar cell producers, different businesses are designing this form of cloth. This will allow men, when in the light, to recharge these tiny gadgets.

3. A recent development in solar panel roofs has come in which the roof is filled with solar panels in most of the houses. The roof of the max in some advanced countries like China Using aluminum or solid alloy to sustain the weight of panels; the homes are constructed of solar panels. These panels produce power for themselves in rural places where grid energy is not feasible and supply energy to neighbors who cannot bear the construction costs. In that area where delivery of electricity is not feasible through other means & transmitting may not be practical due to heavy investment, this will assist surrounding masses and society.
4. In certain metropolitan countries, certain solar energy hobbyists produce electricity, this machine is adequate to fulfill their requirements, and the surplus produced power is supplied to the grid that empowers the national grid. The solar panel should be mounted when appropriate on the cars so that batteries and power charging can be carried out.

CONCLUSION

The solar energy is becoming common and urgency of the day due to the decrease in advisability of natural fuels and the discovery of environmental changes due to traditional method of generation. With the aid of solar energy, other devices in the car can be made usable. This would increase the vehicles' electrical performance whenever the solar rays land on the wall. In Japan and other industrialized countries, where conversion technologies from solar to electricity are commonly and sufficiently utilized, this form of experimentation and usage is also being carried out. It is also concluded that solar power plants should be built in such a manner that they can operate along with hydro and other generation strategies to increase cleanliness Solar panels have also been mounted in the metro railway system in India.

REFERENCES

[1] Rowlands, I. H. Envisaging feed-in tariffs for solar photovoltaic electricity: European lessons for Canada. *Renew Sustain Energy Rev* 2005;9:51-68.

[2] A market for renewable energy credits in the Indian power sector, by Anoop Singh: in *Science Direct Renewable and Sustainable Energy Reviews* 13 (2009): 643-652.

[3] The Value of Grid-Support Photovoltaics in Providing Distribution System Voltage Support, *Proceedings of the American Solar Energy Society's* 1994 Annual Conference.

[4] S. Kannadhasan, M.Isavani and G.Karthikeyan, A Novel Approach Privacy Security Protocol Based SUPM Method in Near Field Communication Technology, *Artifical Intelligence and Evolutionary Algorithms in Engineering Systems* (ICAEES 2014) Organized by Noorul Islam University, Kumaracoil Nagercoil on 23-24 April 2014 Published for Advances in Intelligent Systems and Computing

[5] *Electricity Market Regulations and their Impact on Distributed Generation,* by Thomas Ackermann Goran Andersson Lennart Soder Royal Institute of Technology, Department of Electric Power Engineering Electric Power Systems, Sweden.

[6] Technical, Economical and Regulatory Aspects of Virtual Power Plants T. G. Werner, Germany R. Remberg, Germany, in *DRPT2008* 6-9 April 2008 Nanjing China.

[7] CIGRE Technical Brochure "Grid Integration of Wind Generation"*Working Group 08 of Study Committee C6*, 2009 by H. Weber University of Rostock Germany.

[8] Dahlin, K., Anderegg,W.,Hernandez,R. R., Hiza, N., Johnson, J., Maltais-landry, G. et al., 2011. *Prospects for integrating utility-scale solar photovoltaics and industrial agriculture in the U.S. American Geophysical Unionfallmeeting*, San Francisco, CA. Abstract#B23B-0419.

[9] Eitelhube, G. "Seed Fund Winners - NOMADD: The No-water Mechanical Automated Dusting Device", King Abdullah University of Science and Technology (KAUST), *Energy and sustainability are primary research drivers*, 2012.

[10] S. Kannadhasan and R. Nagarajan, Performance Improvement of H-Shaped Antenna With Zener Diode for Textile Applications, *The Journal of the Textile Institute*, Taylor & Francis Group, DOI: 10.1080/00405000.2021.1944523

[11] Badding, J. "Flexible Silicon Solar-Cell Fabrics May Soon Become Possible", *journal Advanced Materials,* Penn State University, 6 December 2012.

[12] Gielen, D. "Concentrating Solar Power technology", *Energy Technology Systems Analysis Programme, International Renewable Energy Agency*, January 2013.

[13] Rai, G. D. "Non-Conventional Energy Sources", *Khanna Publication,* 4th edition, 2012.

In: Research Challenges in Science ... ISBN: 978-1-68507-008-3
Editors: S. Kannadhasan et al.

Chapter 17

OPTIMIZED ANTIREFLECTION COATINGS FOR HIGHER EFFICIENCY ZNO SOLAR CELL

S. Rajalakshmi[1,*], N. Sundaramurth[2,3,†] and V. Usha[4,‡]

[1]Department of Physics,
University College of Engineering Panruti, Panruti, India
[2]Department of Physics, Annamalai University,
Chidambaram, Tamilnadu, India
[3]Department of Physics, TKG Arts College,
Virudhachalam, Tamilnadu, India
[4]Department of Chemistry,
University College of Engineering Panruti, Panruti, India

ABSTRACT

Dye-sensitized solar cells (DSSCs) have been widely paid attention due to their great potential for low cost energy conversion devices. However, DSSCs show low power conversion efficiency (PCE) because of some efficiency losses, such as reflection of

* Corresponding Author's E-mail: rajalakshmiaup@gmail.com.
† Corresponding Author's E-mail: nagasundaramphysics@gmail.com.
‡ Corresponding Author's E-mail: vushachem@gmail.com.

light by a substrate, charge recombination, carrier trapping in a material and surface power dispersion due to the resistance of cell (material, metal, electrode, etc). Approaches to improve the power conversion efficiency of DSSCs are to reduce these losses. Reduction of light reflection by using three layer anti-reflection coatings is one of the effective approaches. These three layers are expected to reduce the reflection and increase the transmission of light at the glass surface. This leads to an increase of the total flux of photons to photocurrent through the electrode and short circuit current density (Jsc).

1. INTRODUCTION

Solar energy represents the largest energy input into the terrestrial system. Photovoltaic can convert sunlight directly into electricity and so allows creating an independent power supply. Solar cells can generate power in remote areas, because they do not rely on energy sources other than sunlight. In principle, the average sunshine over an area only as large as Wales (UK) would already be enough to supply the world's electricity needs. Photovoltaic thus has a great potential to address the world's energy requirement in the long run, in contrast to conventional resources, which are either finite or which generate pollutants. However, in order to compete with conventional energy technologies, solar cells still need to prove higher power conversion efficiencies and lower costs.

1.1. Nano-Crystal Solar Cell

Nano crystal solar cells are solar cells based on a substrate with a coating of nano crystals. The nano crystals are typically based on silicon, CdTe or CIGS and the substrates are generally silicon or various organic conductors. Quantum dot solar cells are a variant of this approach but take advantage of quantum mechanical effects to extract further performance. Dye-sensitized solar cells are another related approach, but in this case the nano-structuring is part of the substrate. A thin film of nano crystals is obtained by a process known as "spin-coating." This involves placing an amount of the quantum dot solution onto a flat substrate, which is then rotated very quickly. The solution spreads out uniformly, and the substrate is spun until the required thickness is achieved. It is argued that many measurements of the efficiency of the nano crystal solar cell are incorrect and that nano crystal solar cells are not suitable for large scale manufacturing. Recent research has experimented with lead selenide (PbSe) semiconductor, as well as with cadmium telluride photovoltaics (CdTe), which has already been well established in the production of second-generation thin film solar cells.

A dye-sensitized solar cell (DSSCs) is a third generation solar cell which was developed by O'Regan and Gratzell in 1991 (O'Regan and Gratzel, 1991). This inexpensive solar cell (also known as photovoltaic device) works by converting photon from solar energy to electrical energy, based on sensitization of wide band gap semiconductor, dyes and electrolyte (Bisqurt et al., 2004). The functioning of DSSCs is the interlacing of several subsystem which should work in tandem: the dye sensitizer, that is, the dye adsorbed on the semiconductor surface able to absorb the visible and near – IR photons and to pump electrons into the conducting band of semiconductor, the electron-mediator for "hole" conduction and the counter electrode catalytic materials.

The performance of the DSSC is highly dependent on the sensitizer dye and wide band gap material such as TiO2, SnO2, and Nb2O5 (Gratzel, 2003). TiO2 is highly preferable due to its ability to resist the continuous transfer of electron under illumination (in the ultraviolet range). Several studies have addressed the use of SnO2 (Bauer et al., 2002) ZnO (Pradhan et al., 2007). ZnO has great research interest in DSSC due to some fascinating properties. It has higher binding energy (60meV), wide band gap (3.37eV), high break down strength, cohension and exciton stability (Pradhan et al., 2007). Moreover, is one of the hardest materials in the family of II-IV semiconductors. It had been used in DSSC fabrication and achieved up to 0.7% conversion efficiency .

DSSCs are currently the most efficient third-generation solar technology available (O'Regan and Grätzel, 1991). Other thin film technologies are currently certified for small cells at about 11.1% efficiency (Chiba et al., 2006) and several large-scale tests have been conducted to illustrate the promise for commercial application of DSSC concept. (Dai et al., 2008; Toyoda et al., 2004).

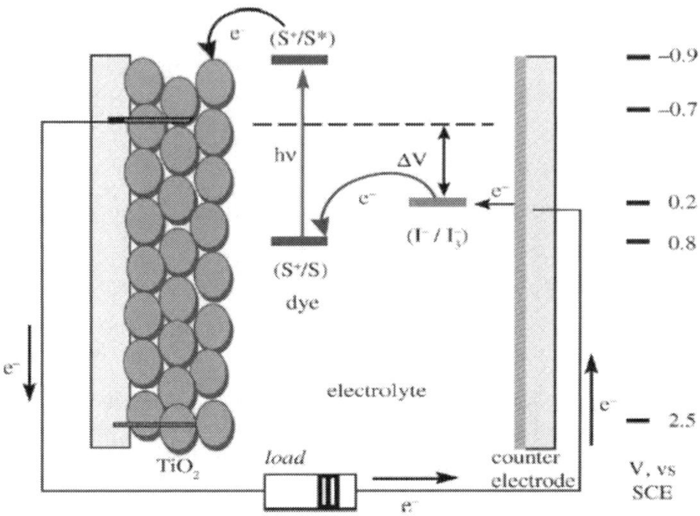

Figure 1. Schematic drawing of a DSC showing the principles of operation.

In the absence of a redox mediator to intercept and rapidly reduce the oxidized dye (S+), recombination with the electrons of the titania layer takes place, without any measurable photocurrent:

$$S^{+} + e^{-}(TiO_2) \rightarrow S \text{ (recombination)} \quad (1)$$

The electrons reaching the counter-electrode through the external circuit reduce in turn the oxidized iodide (Γ) so that the entire sequence of electron transfer reactions involving the dye and the redox mediator (I2-I~) is rendered cyclic:

$$I_3^{-} + 2e^{-} \rightarrow 3I^{-} \text{ (regeneration of I}^{-}\text{)} \quad (2)$$

If cited reactions alone take place, the overall effect of irradiation with sunlight is to drive the electrons through the external circuit, i.e., direct conversion of sunlight to electricity.

1.2. Key Efficiency Parameters of a DSC

The spectral response of the dye-sensitized solar cell depends on the absorption properties of the dye. Characterization of the cell depends on a number of experimentally accessible parameters, including the photocurrent and photopotentials measured under different conditions (open and closed circuit, under monochromatic light or sunlight illumination): 70C, Voc,/sc and Vsc. The term *incident photon-to-electricalconversionefficiency* (IPCE) is a quantum-yield term for the overall charge-injection collection process measured using monochromatic light (single wavelength source). The photocurrent measured under closed circuit I_{sc} is the integrated sum of IPCE measured over the entire solar spectrum:

$$I_{SC} = \int_0^{\infty} IPCE(\lambda).I_{sun}(\lambda)d\lambda \qquad (3)$$

Thus IPCE(lambda)can be expressed as

$$IPCE(\lambda) = 1240\left({}^{I_{SC}}\!\big/_{\lambda_{\phi}} \right) \qquad (4)$$

where lambda is the wavelength, I_{sc} the current at short circuit(mA/cm^2) and Φ is the incident radiative flux(W/m^2).

The overall sunlight-to-electric-power conversion efficiency of a DSC is given by the following expression:

$$\eta = \frac{P_{\max}}{P_{in}} = \frac{I_{SC}.V_{OC}.FF}{P_{in}} \qquad (5)$$

Maximum power obtainable in a photovoltaic device is the product of two terms I_{max} and $V_{max.}$ The value of I_{max} gives the maximum photocurrent obtainable at some "maximum power point." The Fill factor FF is defined as the ratio of ($I_{max}V_{max}/I_{sc}V_{oc}$). The four values I_{sc},V_{oc}, FF and η are the key performance parameter of the solar cell.

The overall efficiency(η_{global}) of the photovoltaic cell can be calculated from the integral photocurrent density(I_{ph}), the open- circuit photovoltage (V_{oc}), the fill factor of the cell (FF) and the intensity of the incident light(I_s = 1000 W/m_2)

$$\eta_{global} = (I_{ph}.V_{oc}.FF) / I_S \tag{6}$$

The measured photocurrent will depend on the light intensity; the efficiency of charge injection in the excited-state quenching process; the degree of recombinationof electrons with the oxidized dye (S+); and the efficiency of charge transport inthe titania films to the counter-electrodes. Maximum photovoltage obtainable in such sensitized solar cells is the energy gap between the chemical potential level of the mediating redox electrolyte and the conduction band level of Ti02.

Figure 4 shows representative photocurrent-voltage data for a dye-sensitized solar cell based on the Ru-bpy sensitizer N945 measured under AM 1.5 sunlight illumination (100 mW/cm-2) (red line), along with the dependence of the power conversion efficiency for monochromatic light (blue line). The cell active area is 0.158 cm2. The inset shows a photocurrent action spectrum. For efficient DSCs with overall conversion efficiency >10%, at 1 sun (AM 1.5) irradiation, typical photocurrent is of the order of 15-20 mA/cm2, the photovoltage in the range 650-750 mV, with a fill factor of 0.65-0.80.

The overall *sunlight-to-electric-power conversion efficiency* of a DSC can be expressed as the product of three key terms:

$$H = \eta_{abs}.\eta_{inj}.\eta_{coll} \tag{7}$$

where η_{abs} is the efficiency of light absorption by the dye; η_{inj} is the efficiency of charge injection from the excited state of the dye;and η_{coll} is the efficiency of charge collection in the mesoporous oxide layer. An ideal photosentisizer will be the.

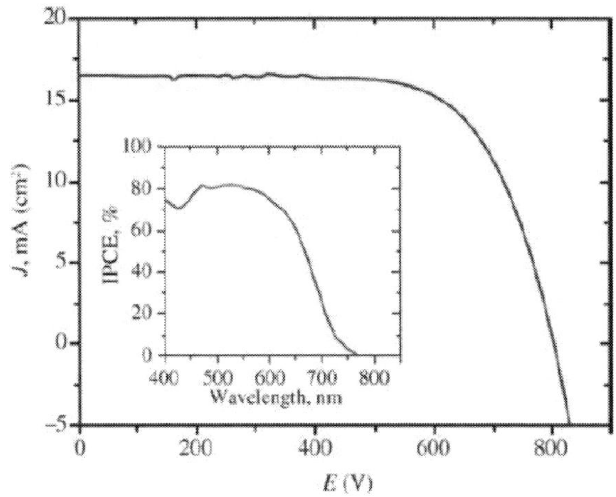

Figure 2. Photocurrent-voltage curve of a solar cell based on the Ru-bpy sensitizer N945measured under AM 1.5 sunlight illumination (100 mW cm-2) (red line).

The cell active area is 0.158 cm2. Insert in blue shows the monochromatic wavelength dependence of the photocurrent. One that absorbs all sunlight in the visible-near IR region with high absorption cross section (coefficient). The efficiency of charge injection T7inj depends on the number of low-lying electronic excited states below the conduction-band edge of the oxide semiconductor and the ability of these states to undergo electron transfer with the titania in preference to other decay channels of the excited state. For efficient electron transfer with the titania, a good electronic coupling between the electron-acceptor level of titania and the highest occupied molecular orbitals (HOMO) of the dye is required. As has been shown for many photoredox reactions of molecules and coordination compounds in solution, the energetics of the charge-injection step can be determined by considering the redox potentials of the dye excited state relative to the acceptor (conduction band) level of the oxide substrate. A moderate driving force of approximately 200 mV ensures that the excited-state electron transfer occurs rapidly and quantitatively.

1.3. Key Components of the DSC

Dye-sensitized solar cells have many components that have to be optimized, that includes substrate glass with the transparent conducting oxide TCO layer; mesoporous titania Ti02 layer; underlayer(s); dye; electrolyte solvent; redox mediator; and a counter-electrode. In the case of solar modules, different modes of interconnection of individual cells must be taken into account to optimize the active area for power generation.

1.4. Substrates for the DSC

As mentioned earlier, the DSC has a sandwich structure involving two transparent conducting oxide-glass substrates. The requirements for the TCO substrate are low sheet resistance (nearly temperature independent to the high temperatures used for sintering of the $Ti0_2$ layer, 450-500°C); and a high transparency to solar radiation in the visible-IR region. Typical sheet resistance of the TCO used is 5-15 Ω/square. The cost of TCOs rises steeply with lower sheet resistance and better light transmittance. The cost of these two substrates of the electrodes is nearly half of the total cost of the solar cell. Both indium-doped tin oxide ($In:Sn0_2$, ITO) and fluorine doped tin oxide ($F:Sn0_2$, FTO) have been employed. Use of glass substrates confers good protection against penetration of oxygen or water. But the heavy weight of the glass renders this form of DSC non-portable, restricting its uses to terrestrial power generation. Indium-doped tin-oxide TCO is the most common substrate used in many photonic and optoelectronic devices. This justifies its mass-production on the industrial scale. Unfortunately it has been found that the thermal stability of the ITO glass is not good, with layers peeling off the glass and/or formation of defect sites on the surface that can reduce the solar conversion efficiency of the DSC. Hence, the currently preferred TCO for DSC application is fluorine-doped tin oxide ($F:Sn0_2$). There have been several studies directed towards improving the efficiency-loss problems related to the usage of ITO for DSCs.

In order to obtain optimal adhesive and rheological properties of a mesoporous film with the TCO substrate, suitable binders are added to the colloidal solutions of Ti02 prior to film deposition by doctor-blade techniques or screen printing. Sintering of the oxide layer at 450-500°C gives the film two important properties: the sintering brings the individual colloidal particles to come into close contact so that the conductance and charge collection properties of the titania layer are improved; and the aerial oxidation at elevated temperature removes all the organic matter (potential trap sites) from the mesoporous film. There have been several studies on the performance of titania layers deposited using different precursors and coating procedures. Conducting oxide layers can be deposited on a wide variety of substrates, including polymer-based plastics. Advantages of such DSCs are low weight, flexibility and preparation protocols that are amenable to established industrial methods such as roll-to-roll printing.

A third attractive substrate is thin metal foils, such as titanium or stainless steel. They have essentially same advantages as the polymer-based and plastic substrates. With metal substrates, care must be taken against corrosion of the metal by the electrolyte.

1.5. $Ti0_2$ as the Photoelectrode

Many wide band gap oxide semiconductors (Ti02, ZnO, Sn02, ...) have been examined as potential electron acceptors for DSCs. Ti02 turned out to the most versatile, delivering the highest solar-conversion efficiency. Ti02 is chemically stable, nontoxic and readily available in vast quantities. It is the basic component of white paints. Particulate dispersions of Ti02, known as P25, is produced by Degussa by the ton for the paint industry. Thousands of publications have appeared on the preparation of colloidal particles of titania by the sol-gel hydrolysis route using different precursors. An important requirement for the semiconductor is high transport mobility of the charge carrier to reduce the electron-transport resistance. ZnO, Sn02, Nb205 and many titanates of the general formula MT1O3 have been examined as alternative oxides for the DSCs. Some of these studies are reviewed in of this volume.

TiO_2 has many crystalline forms, with anatase, rutile and brookite being the easily accessible ones. Rutile has a slightly lower band gap as compared to anatase and can absorb a few percent of sunlight in the near-UV region. In the standard version of DSCs, typical film thickness is 2-15 /mi, and the films are deposited using nanosized particles of 10-30 nm. The highest solar-conversion efficiency is obtained in double layer structures, where an under layer of thickness 2-4 /mi is first deposited using larger (200-300 nm) size particles. While the beneficial effects of the under layer has been unambiguously established for the quasi-solid-state version of DSC that uses hole transport materials (HTM), clear trends of beneficial effects have yet to be observed for the common liquid-electrolyte-based DSCs. A number of experimental approaches have been used for the deposition of the mesoporous film on the TCO substrate. The most commonly used are doctor-blading, screen printing and spray-drying.

2. Procedure for the Preparation of DSCs

For the key anode component, a mesoporous film of TiO_2 (of several micron thickness) is deposited using colloidal particles prepared by the controlled sol-gel hydrolysis of Ti-alkoxides. This electrode acts as a sponge and can readily take up a variety of organic and inorganic dye molecules. The dye is deposited by immersion in a stock solution for less than an hour. Depending on the method of preparation and material processing, the oxide layer can be highly translucent (ideal for integration as part of the building architecture) or opaque. The mesoporous layer consists of well interconnected colloidal particles in the size range of 15-30 nm, and the layer thickness is in the range of 5 to 15 /mi. Depending on the particle size, the effective surface area (for dye adsorption), porosity and pore volume (for penetration of the redox electrolytes) of the oxide layer can vary significantly. For the best photovoltaic performance, post treatment of the mesoporous oxide layers with TiCl4 is also applied and enhanced haze obtained by optimization of the light scattering properties of the mesoporous layer.

2.1. Dyes for DSC Photosensitizers

Along with the mesoporous oxide layer, a key component of the DSC is the photosensitizer ("dye") that absorbs the solar radiation and injects electrons into the conduction band of the oxide substrate. Thousands of small-, medium- and large-sized molecules have been examined as potential candidates during the past two decades. In addition to organic molecules, coordination complexes of transition metals with polypyridine or porphine ligands have been studied.

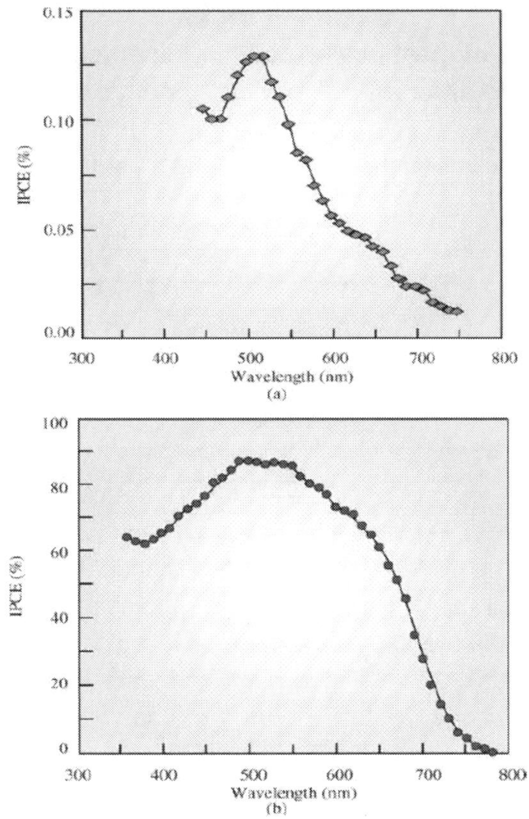

Figure 3. A comparsion of the monochromatic power conversion efficiency of the same Ru-dye N719 deposited (a) on the single crystal TiO_2 (anatase) electrode and (b) on a mesoporous oxide based DSC.

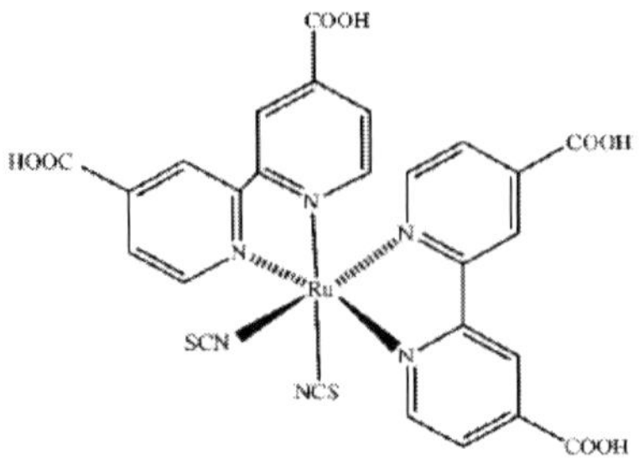

N-3 "red" dye

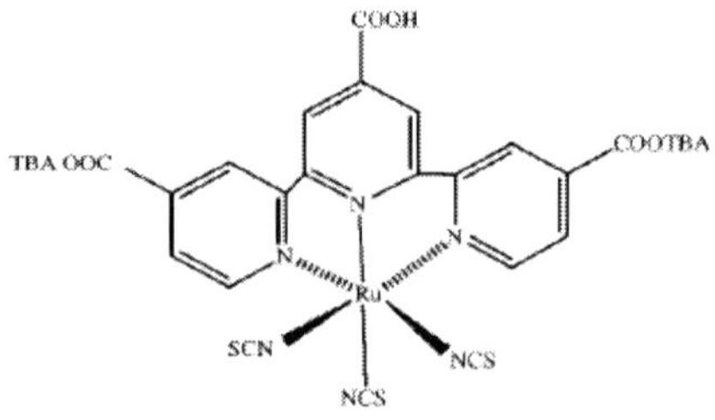

749 "black" dye

Figure 4. Chemical drawings of "red" and "block" dye used in DSCs.

Table 1. The highest power-conversion efficiencies obtained using different polypyridine complexes of Ru

Dye	Surface area(cm^2)	η(%)	V_{oc}(V)	I_{SC}(mA/cm^2)	FF(%)
N-719	<1	11.2	0.84	17.73	74
N-749	0.219	11.1	0.736	20.9	72
N-749	1.004	10.4	0.72	21.8	65
N-719	1.31	10.1	0.82	17.0	72
N-3	2.36	8.2	0.726	15.8	71
K-19	<1	9.19	0.768	16.4	73
N-945	<1	9.55	0.759	17.25	73
N-621	<1	8.69	0.766	16.22	70
Z-907	<1	7.18	0.70	15.09	68
J-13	<1	7.83	0.703	15.65	71

With reference to the simplified DSC model, it is possible to list a number of desirable properties for a photosensitizer: strong light absorption in the visible and near- IR region (for efficient light harvesting); good solubility in organic solvents (for facile deposition from stock solutions in few hours or less); presence of suitable peripheral anchoring ligands such as -COOH (to promote the effective interaction of the dye with the oxide surface and thus the coupling of donor and acceptor levels); suitable disposition of the HOMO and LUMO of the dye molecule.

In standard DSCs using mesoporous TiO_2layers and iodide-triiodide redox electrodecharges from the electronically excited state);

good thermal stability; and good chemical stability (to retain the chemical identity over repetitive oxido-reduction cycles).

Examples of dyes that function well in DSCs are coordination complexes of Ru and Os squaraines, porphyrins, perylenes, pentacenecyanines and coumarins. The paradigm here is the polypyridine complexes of Ru(II) with one or more carboxylic acid groups as a peripheral substituent. Because of the complex chemical nomenclature, photosensitizers are most often referred to by their trivial names (for example, N-3 and N-749 are commonly also called "red" and "black" dyes, shoThese red and black dyes are the benchmark complexes used as standards for the development of other dyes and components of the DSC. Solar conversion efficiency of over 10% for labcells have been obtained with many Ru dyes of polypyridine family (see data presented in Table 1 for various Ru-polypyridine complexes developed at the EPFL in Lausanne). Introduction of long alkyl chains as a substituent in the polypyridine ring increases the hydrophobicity of the dyes significantly, thereby reducing the stability. Chemical drawings of "red" and "black" dyes used in DSCs. Photochemical and photo electrochemical approaches 27 problems arising from residual water content in the oxide layers. The so-called Z-, K- and C-series of these dyes are typical examples.

Extension of the 7r-conjugation of polypyridine groups with additional antennatype substituent groups leads to substantial increase in the visible-light-absorption properties and higher molar absorbance. The C101, C103 and Z991 dyes with thiophene- based antenna units illustrate this enhancement process. The metal-to-ligand charge transfer (MLCT) band in the visible region of these complexes is red shifted from 530 to 553 nm. The molar extinction coefficient is increased from 14 500 to 21 200 M_1 cnr1, allowing the use of thinner titania films. With IPCE values of >70% obtainable up to 800 nm, this new generation of dyes permit thefabrication of DSCs with solar-light-conversion efficiencies well over 10%. Other examples of D-π-A sensitizers are C202, C206, C207, C208 dyes with different triarylamine electron donors in conjugation with a thienothiophene unit as a bridge and a cyanoacrylic acid acceptor. The J-series of dyes that include triaryl groups, such as J-13 and J-16, show similar improvements. Many organic dyes with an extended π-system of the donor-acceptor type also have been found to work efficiently in DSCs. Encouraging results, with solar-conversion efficiencies in the range of 6-8%, have been reported using purely "organic" sensitizers with bis-dimethylfluorenylamino as the donor and a dithiophene unit as a spacer. Typical examples of this JK-series are JK2, JK45 and JK46. With indoline dye D205 and C217, solar conversion efficiency of 9.5% and 9.8% respectively have been obtained. Imahori, Bauerle and coworkers have reviewed some of the exciting developments in this area.

3. Electrolytes for the DSC

The electrolyte plays a very important role in the DSC by facilitating the transport of charge between the working and counter electrodes (WE and CE). The ideal electrolyte solvent is one that has very low viscosity, negligible vapor pressure, high boiling point and high dielectric properties. From industrial perspective, factors such as robustness (chemical inertness), environmental sustainability, and easy processing are also important.

Among the many redox mediators examined, the iodide/triiodide couple has been found to be the best. Under optimized conditions, the iodide has a concentration of a few millimolar (and iodine ten times greater), dissolved in a mixture of acetonitrile: valeronitrile (ACN:VN) or neat N-methoxypriopionitrile MPN.

Ionic liquids have many desirable properties, such as high thermal stability, negligible vapor pressure, non-flammability and excellent environmental compatibility. The limitation, however, is their high viscosity, which is one to two orders of magnitude higher than the volatile solvents. This reduces significantly the diffusion-limited current, with the associated overall reduction in the power-conversion efficiency. One approach overcome this problem is to mix low-viscosity solvents with ionic liquids. Table 2.1 shows the typical composition of two widely used electrolytes for DSC studies.

4. The Counter Electrode

To balance the charge and regenerate the key components, the oxidized form of the mediator needs to be reduced by the electrons flowing through the external circuit.

Table 2. Typical currently employed solvent-based electrolyte compositions

Electrolyte type	High η	Robust
Solvent	ACN:VN(3/1)	MPN
Iodide compound	1 M PMII	1 M PMII
Iodine	$0.03MI_2$	$0.15MI_2$
Additive 1	0.1 M GuSCN	0.1 M GuSCN
Additive 2	0.5 M TBP	0.5 M NBB
Efficiency	10.5% (<1cm^2)	9% (<1cm^2)

ACN = acetonitrile, VN = valeronitrile, MPN = 3-methoxypropionitrile, PMII = l-propyl-3-methylimidazolium iodide, GuSCN = guanidinium thicocyanate, TBP = tert-butylpyridine, NBB = 1-butyl-1-benzimidazole.

Passing through the counter electrode. For iodide-I_2 mixture, the oxidized form corresponds to triiodide and reduction involves 2 electrons.

$$I_3^- + 2e^- \rightarrow 3I^-$$

To reduce losses, the counter-electrode material should show good electro catalytic properties. The material of choice has been platinum. It suffices to have very low quantities of platinum deposited onto a TCO (a few /ig/cm2) as finely divided particles. As an electrode material, Pt shows excellent chemical stability and very low overpotential for the tri-iodide reduction reaction.

4.1. Effect of the Thickness of ZnO Photo Electrode in DSSC

Solar energy is one of the promising renewable energy resources. Solar cell or photovoltaic cell is a device converting solar energy into electricity. It was made of silicon in the first development. In 1991 Dye-Sensitized Solar Cell (DSSC), was invented by Grätzel which utilized Ru-dye sensitizer and oxide semiconductor. In contrast to Si-based solar cell, solar energy is absorbed by dye molecules and charge separation is gained through wide band gap of oxide semiconductor. TiO_2-based DSSC has been reported to produce the highest solar to electric energy conversion efficiency up to 11%. TiO_2 is used in DSSC applications since it has high photo activity. Another photoactive oxide semiconductor, ZnO, is interesting to be employed as DSSC photo electrode. ZnO can be easily synthesized in various nanostructures. It has higher band gap energy and bulk electron mobility than that of TiO_2. It allows better photo electro chemical properties of DSSC. However, conversion efficiency of 5% is the best record for ZnO-based DSSC. Thickness of ZnO photo electrode is also an important of factor to change the efficiency of DSSC. Therefore the effect of ZnO- photo electrode thickness is investigated in this work [1].

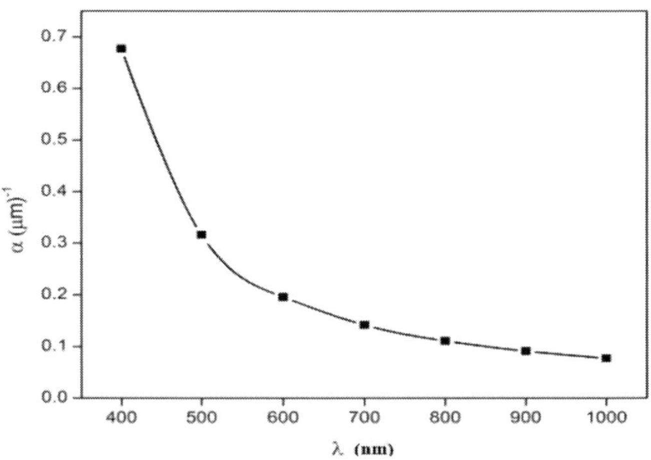

Figure 5. Absorption coefficient of ZnO material vs Wavelength.

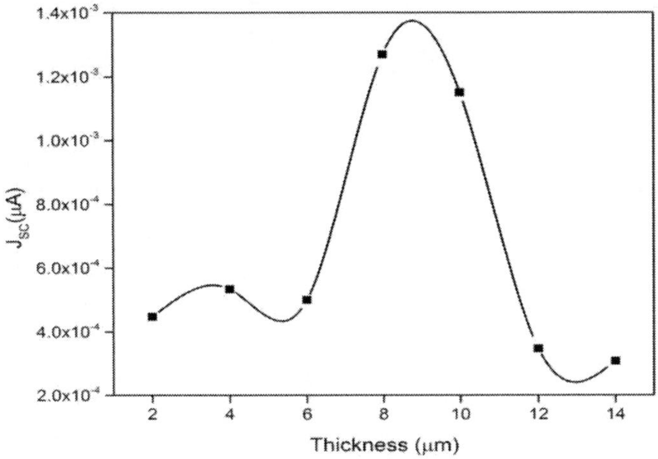

Figure 6. Variation of Short circuit current with the thickness of ZnO material.

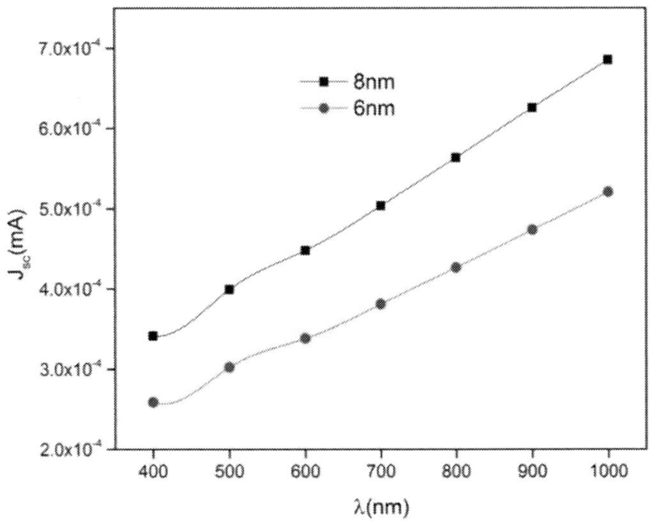

Figure 7. Variation of Short circuit current with the wavelength of light.

5. Result and Discussion

The performance of front-illuminated DSSC is largely dependent on ZnO film thickness. A part of the incident light which is transmitted through the electrolyte is absorbed by the electrolyte in the case of back-illuminated DSSCs. There, it is worthy to investigate the effect of film thickness which affects the performance of DSSCs. The J_{SC} is calculated with ZnO film thickness from 2 to 14nm. The film thickness of ZnO had a significant influence on short circuit current. The variation of absorption coefficient with the electromagnetic radiation of wavelength 400nm-1000nm is studied and is plotted in Figure 5. From the graph it is found that absorption coefficient decrease with wavelength and become constant at longer wavelength region. Figure 6 shows that the effect of film thickness on short circuit current of DSSCs using with N719 Dye. The short circuit current increases and gets a maximum value at 8nm of the film thickness and decrease with thickness. This decrease in the short circuit current is generally due to the higher charge recombination, with more limited mass transport in the thicker photoanodes. The changes of FF (Filling Factor) and V_{OC} (Open circuit Voltage) corresponding to changes in the thickness of the photoelectrode are less significant than the change of short circuit current. The short circuit current can also be studied in terms of wavelength of light. Figure 7 shows this variation of short circuit current for two different thickness of ZnO Nanoparticles. It is found that the short circuit current increase with wavelength of light. Since efficiency of DSSC depends on the short circuit current, these types of studying as short circuit current may helpful to improve the efficiency of DSSCS.

5.1. Efficient Anti-Reflective Coating in ZnO DSSC

Dye-sensitized solar cells (DSSCs) have been widely paid attention due to their great potential for low cost energy conversion devices. However, DSSCs show low power conversion efficiency (PCE) because of some efficiency losses, such as reflection of light by a substrate, charge recombination, carrier trapping in a material and surface power dispersion due to the resistance of cell (material, metal, electrode, etc). Approaches to improve the power conversion efficiency of DSSCs are to reduce these losses. Reduction of light reflection by using three layer anti-reflection coatings is one of the effective approaches. Three layers are expected to reduce the reflection and increase the transmission of light at the glass surface. This leads to an increase of the total flux of photons to photocurrent through the electrode and short circuit current density (Jsc), [2-13].

In this work the three layers anti reflecting coating is used to reduce the reflectivity loss. The three layers ARC coating materials in the system are MgF_2/ZnSe,/ZrO2. ARC system is calculated the reflectivity of the theoretically using the following equation [14-15].

$n_{4(Air)}$	n_3	n_2	n_1	$n_{0(ZnO)}$

Figure 8. The Schematic structure of Multilayer ARC Coating.

$$R_1 \exp\left[-i\rho_1(\lambda)\right] = r_1 + r_0 \exp(-2ikg_1) \tag{8}$$

$$R_3 \exp\left[-i\rho_3(\lambda)\right] = r_3 + r_4 \exp\left(-2ikg_3\right) \tag{9}$$

$$\left(R_3\right)^2 = r_3^2 + r_4^2 + 2r_3r_4 \cos 2kg_3 \tag{10}$$

$$\rho_1(\lambda) + \rho_3(\lambda) + 2kg_2 = 2\pi N \tag{11}$$

λ_O = reference wavelength

N is a positive integer.

where

$K = \frac{\lambda_O}{\lambda}$

λ_0 is the centre wavelength

g_i = Optical thickness

h_i is the metrical thickness of the ith film

$R_1 = R3$ Amplitude reflection coefficient of the film

$$r_1 = \frac{n_2 - n_1}{n_2 + n_1} \quad r_0 = \frac{n_1 - n_0}{n_1 + n_0} \quad r_4 = \frac{n_3 - n_4}{n_3 + n_4} \quad r_3 = \frac{n_2 - n_3}{n_2 + n_3}$$

n_0, n_1, n_3, n_4 = Refractive index of anti reflective coating 1, 2, 3 and substrate

r_1, r_0,, r_4, r_3=Reflectivity of medium n_1 to n_2, n_1 to n_0, n_3 to n_4, and n_3 to n_4

The reflectivity of ZnO is calculated from [25]

$$R = \frac{(n-1)^2}{(n+1)^2}$$

It is necessary to express the reflectances in terms of the physical constants of the system.

It can be shown that the three-layer dielectric coating has zero reflectance when

$$R_1 = R3 \ where \quad g_i = \frac{2\pi n_i h_i}{\lambda_0}$$

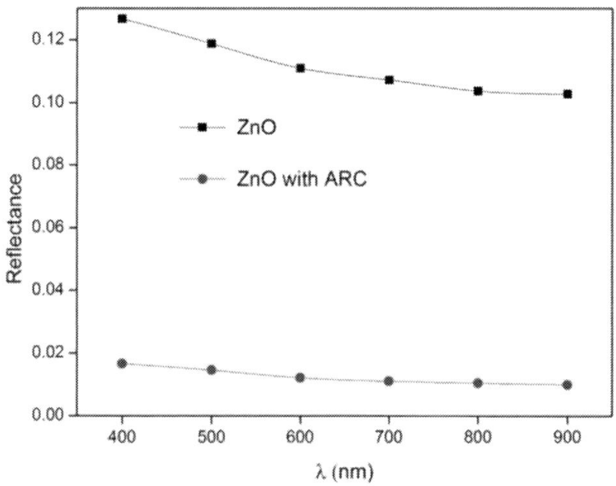

Figure 9. The variation of reflectance with wave length.

The influence of system is the three layers ARC. Reducing the reflectivity in dye synthesis solar cell is analysed. The Table 3 shows that the reflectivity variation of the system with and without the three layers ARC of ZnO. Substrate for a wavelength range of 400 -900nm. The reflectance of DSSC for various wavelength of light with or without

ARC is studied as is shown in Figure 9. It is found that the reflectivity is very much reduced when the three layers ARC is incorporated in the DSSC. The reflectivity also affects the short circuit current in DSSC. The variation of short circuit current with wavelength for keeping thickness of ZnO at 8nm and 6nm are studied and this plotted Figure 10 and 11. In both the cases, we found similar behaviour and it is notice that DSSC with ARC provide higher short circuit current compared to short circuit current without ARC. The improvement of short-circuit current enhanced to increase the efficiency ZnO DSSC solar cell [16].

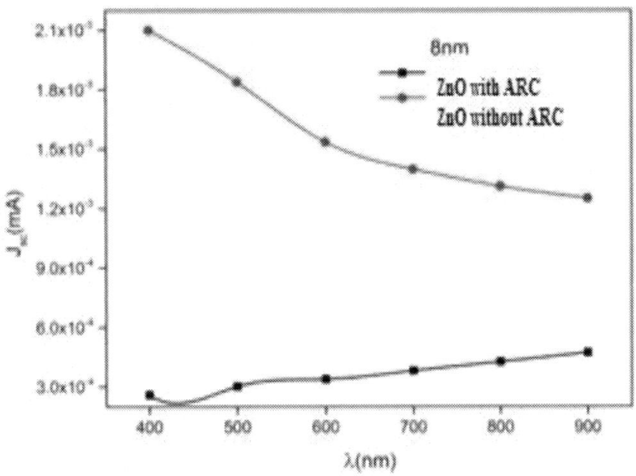

Figure 10. The variation of reflectance with wavelength at 8nm thickness.

Table 3. The Reflectivity Analysis of ZnO DSSC with and without ARC

λ(nm)	Reflectivity of ZnO	Reflectivity of ZnO with ARC
400	0.12671	0.01665
500	0.11875	0.0146
600	0.11095	0.01221
700	0.1072	0.01111
800	0.10368	0.01043
900	0.10271	0.00996

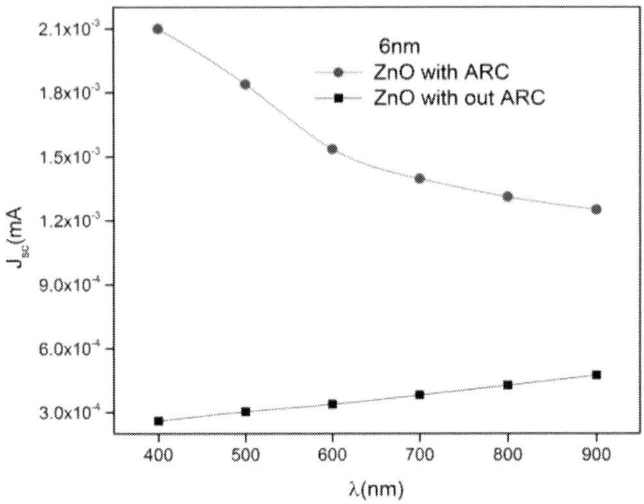

Figure 11. The variation of reflectance with wavelength at 6nm thickness.

Conclusion

A detailed study of Dye sensitised solar cell, its working processes and the role of internal parameters in calculating the efficiency are analysed. The importance of the three layers Antireflective coating system in ZnO based Dye sensitised solar cell are studied. It is bound that the performance of Dye sensitised solar cell is largely dependent on ZnO film thickness.

The short circuit current is one of the important parameters to increase the efficiency of Solar cell. The film thickness of ZnO had a significant role on J_{SC}. By varying the thickness of the ZnO substrate film from 2 to 14nm, the short circuit current is analysed. Hence the reflectance of 3 layers ARC which ZnO substrate is studied.

The ARC plays an important role in increasing the efficiency of solar cell. The reflectance of Dye sensitised solar cell with or without ARC is analysed. It is concluded that with ARC the reflectance is very much reduced. Hence this may be used to increase efficiency of Dye sensitised solar cell.

REFERENCES

[1] Irimpan L., V. P. N. Nampoori, P. Radhakrishnan, *Spectral and nonlinear optical characteristics of nanocomposites of ZnO – Ag,* 455 (2008) 265–269. doi:10.1016/j.cplett.2008.02.097.

[2] Tripathi B., P. Yadav, M. Kumar, Theoretical upper limit of short-circuit current density of TiO2 nanorod based dye-sensitized solar cell, *Results Phys.* 3 (2013) 182–186. doi:10.1016/j.rinp.2013.09.002.

[3] Kalyanasundaram E. K., M. Bertoz, J. Bisquert, F. De Angelis, H. Desilvestro, F. Fabregat-santiago, S. Fantacci, A. Hagfeldt, S. Ito, K. Jiang, K. Kalyanasundaram, P. V Kamat, L. Kavan, J. Moser, K. Nazeeruddin, L. Peter, H. J. Snaith, G. Tulloch, S. Tulloch, S. Uchida, S. Yanagida, J. Yum, *Dye-Sensitized Solar Cells*, 2010. doi:10.1021/cr900356p.

[4] Lau S. C. T., J. Dayou, C. S. Sipaut, R. F. Mansa, Development in photoanode materials for high efficiency dye sensitized solar cells, *Int. J. Renew. Energy Res.* 4 (2014) 665–674. http://www.scopus.com/inward/record.url?eid=2-s2.0-84928015097&partnerID=tZOtx3y1.

[5] Li Y., S. H. Ku, S. M. Chen, M. A. Ali, F. M. A. AlHemaid, Photoelectrochemistry for red cabbage extract as natural dye to develop a dye-sensitized solar cells, *Int. J. Electrochem.* Sci. 8 (2013) 1237–1245.

[6] Zhou Y., Q. He, Y. Yang, H. Zhong, C. He, G. Sang, W. Liu, C. Yang, F. Bai, Y. Li, Binaphthyl-containing green- And red-emitting molecules for solution-processable organic light-Emitting diodes, *Adv. Funct. Mater.* 18 (2008) 3299–3306. doi:10.1002/adfm.200800375.

[7] Irimpan L., V. P. N. Nampoori, P. *Radhakrishnan, Spectral and nonlinear optical characteristics of nanocomposites of ZnO – Ag,* 455 (2008) 265–269. doi:10.1016/j.cplett.2008.02.097.

[8] El Tayyan A. A., Dye sensitized solar cell: parameters calculation and model integration, *J. Electron Devices.* 11 (2011) 616–624.

[9] Hassanien A. S., A. A. Akl, Influence of composition on optical and dispersion parameters of thermally evaporated non-

crystalline thin films, *J. Alloys Compd.* 648 (2015) 280–290. doi:10.1016/j. jallcom.2015.06.231.

[10] Soudi A., P. Dhakal, Y. Gu, Diameter dependence of the minority carrier diffusion length in individual ZnO nanowires, *Appl. Phys. Lett.* 96 (2010) 2008–2011. doi:10.1063/1.3456390.

[11] Mansour A. F., S. F. Mansour, M. A. Abdo, *Improvement Structural and Optical Properties of ZnO / PVA Nanocomposites*, 7 (2015) 60–69. doi:10.9790/4861-07226069.

[12] Singh C., S. Hemant, K. Singh, *Green Energy and Technology Anti-reflection and Light Trapping in c-Si Solar Cells*, Springer;, 2017.

[13] Anggraini P. N., L. Retnaningsih, L. Muliani, J. Hidayat, Optimization of TiO2 photoelectrode with titanium isopropoxide for flexible dye sensitized solar cells, *AIP Conf. Proc.* 1729 (2016). doi:10.1063/1.4946918.

[14] Priyadarshini B. Geetha, A. K. Sharma, Design of multi-layer anti-reflection coating for terrestrial solar panel glass, *Bull. Mater. Sci.* 39 (2016) 683–689. doi:10.1007/s12034-016-1195-x.

[15] Smith S. D., Design of Multilayer Filters by Considering Two Effective Interfaces, *J. Opt. Soc. Am.* 48 (1958).

In: Research Challenges in Science … ISBN: 978-1-68507-008-3
Editors: S. Kannadhasan et al.

Chapter 18

STRUCTURAL PROPERTIES OF FE DOPED ZNO NANOPARTICLES

S. Rajalakshmi[1,*], N. Sundaramurth[2,3,†] and V. Usha[4,‡]

[1]Department of Physics,
University College of Engineering Panruti, Panruti, India
[2]Department of Physics, Annamalai University,
Chidambaram, Tamilnadu, India
[3]Department of Physics, TKG Arts College,
Virudhachalam, Tamilnadu, India
[4]Department of Chemistry,
University College of Engineering Panruti, Panruti, India

ABSTRACT

The brief introduction to nanotechnology, it is important to discuss about the Ferric nano materials and its application of various fields. Moreover, the highlights of Fe doped ZnO nanoparticles and its applications were explained under suitable

* Corresponding Author's E-mail: rajalakshmiaup@gmail.com.
† Corresponding Author's E-mail: nagasundaramphysics@gmail.com.
‡ Corresponding Author's E-mail: vushachem@gmail.com.

conditions. The synthesis method adopted for the preparation of ZnO and Fe doped ZnO nanoparticles are discussed. The detailed experimental techniques used to characterize the synthesized Fe doped ZnO nanoparticles are explained well and the results elaborate the various studies such as XRD, FTIR, VSM, which have been carried out of synthesized nanoparticles.

1. INTRODUCTION

1.1. Introduction of Nanotechnology

In recent years, nanotechnology has encouraged enormous scientific modern development in the field of research and technology. The study of nanotechnology depends upon the application of little things which are able to use across all fields such as chemistry, biology, physics, material science, and engineering. A nanoparticle is core atoms which perform as a complete unit in terms of transport and possessions [1]. Nano means a billionth or 10^{-9} m. Its nano particle size range from (1-100 nm) owing to small size it live in a position in dissimilar fields of nanoscience and nanotechnology. Nano size particles are unique in nature because the nano size increase the surface to volume ratio and also its physical, chemical and biological properties are diverse from bulk material, so the major aim to study its minute size is to trigger chemical activity with distinct crystallography that increases the surface area [2-4]. In current years much research is going on metallic nanoparticles and its properties like a data storage capacity [5]. Quantum size effects and the large surface area of magnetic particles dramatically change contributions and spin canting.

The nano magnetic materials were synthesized by varying grain size with the different techniques like mechanical milling, sol gel method, laser ablation method, oxidation method, reverse micelle polyol process and chemical precipitation method etc. The remarkable properties of the nanomagnetic materials are attracted in understanding the effect of particle grain size on the magnetic properties of both soft and hard magnetic materials.

Table 1. Classification of Ferrite material

S. No	Parameter	Garnet	Spinel	Hexagonal
1	Structure	Cubic	Cubic	Cubic
2	Molecular formula	$M_3(Fe_5O_{12})$	$M(Fe_2O_4)0_3$	$M(Fe_{12}O_{19})$
3	Space Group	la-3d	Fd3m	$P6_3$/mmc, D6h
4	Interstitial Sites	a, c and d	(A) & [B]	12k,2_a,$4f_2$,$4f_1$, 2b
5	Examples	$Y_3Fe_5O_{12}$	$CoFe_2O_4$	$CaFe_{12}O_{19}$
6	Lattice Constant (Å)	12.37	8.38	5.83
7	Curie Temp. (K)	553	668	673
8	Saturation. emu/gm Magn	26	63	66
9	Electrical Resistivity, Ω-cm	1×10^{14}	1×10^8	1×10^6

1.1.1. Ferrite

Ferrite materials have great interest with the general chemical formula of AB_2O_4, where A and B stands for various metal cations; generally with iron. Ferrite materials are a class of spinals i.e., materials that adopt a crystal motif consisting of cubic closed pack oxides with A cations occupying $1/8^{th}$ Of the octahedral voids and B cations occupying half of the octahedral voids. For inverse spinal structure, half the B cations occupy tetrahedral sites and both the A and B cations occupy the octahedral sites. Divalent, trivalent and quadric-valent cations can occupy the A and B sites and they include Mg, Zn, Fe, Mn, Al, Cl, Ti and Si.

1.1.2. Zinc Ferrite

Zinc ferrite ($ZnFe_2o_4$) has normal spinel structure it is commercially important material and has been widely used in magnetic applications, gas sensors, catalysts, photo-catalysts and absorbent materials because of its excellent electrical and magnetic properties. Zinc is known to play a decisive role in determining the ferrite properties hence the compositional variation in the ferrites has been brought about by varying the zinc content. The redistribution of metal ions over the tetrahedral and octahedral sites in the spinel lattice on incorporation of zinc is responsible for the modification in ferrite properties. Among the spinel ferrites, zinc ferrite has been the subject of studying because of its unique properties such has chemical and thermal stability and the particle size dependence of magnetic properties. Bulk $ZnFe_2\,o_4$ has a normal spinel structure with the Zn^{2+} ions without magnetic moment in the tetrahedral sites, which makes anti ferromagnetic behavior. However, it has been demonstrated that both the crystallite size and synthesis method can affect the magnetic properties of nanocrystalline zinc ferrite. For example, there distribution of iron and zinc cations into octahedral and tetrahedral sites resulted in ferromagnetic behavior in zinc ferrite nanoparticles. Spinel zinc ferrite is a semiconductor that has potential application in the conversion of solar energy from sunlight. However, because of the lower valence band potential and property in photoelectric conversion, zinc ferrite cannot be used directly in the photo catalytic destruction of toxic organic compounds.

1.1.2.1. Application of Zinc Ferrite

It is principally used as inductive components in a huge multiplicity of electronic circuits such as low-noise amplifiers, filters voltage-controlled oscillators, impedance matching networks, for instance, the recent applications as inductors obey, among other tendencies, to the general trend of miniaturization and integration as ferrite multilayer's for passive functional electronic devices. The multilayer technology has become a key technology for mass production of integrated devices; multilayer allows a high degree of integration density. The high-frequency applications such as telecommunications and radar systems, as microwave technology require higher frequencies and bandwidths up to 100 GHz.

1.1.3. Objective of the Present Work

- The main focus of this thesis is to synthesize the un-doped ZnO and Fe doped ZnO nanostructures via the Sol-gel method.
- The structure and particle size of the Pure and Fe doped ZnO are characterized by the X-ray diffraction.
- The functional groups of synthesized ZnO and Fe doped ZnO nanoparticles are analyzed by the FTIR spectra.
- Especially, the magnetic properties of Pure and Fe doped ZnO are studied using Vibrating Sample Magnetometer.

2. MATERIALS AND SYNTHESIS PROCEDURE

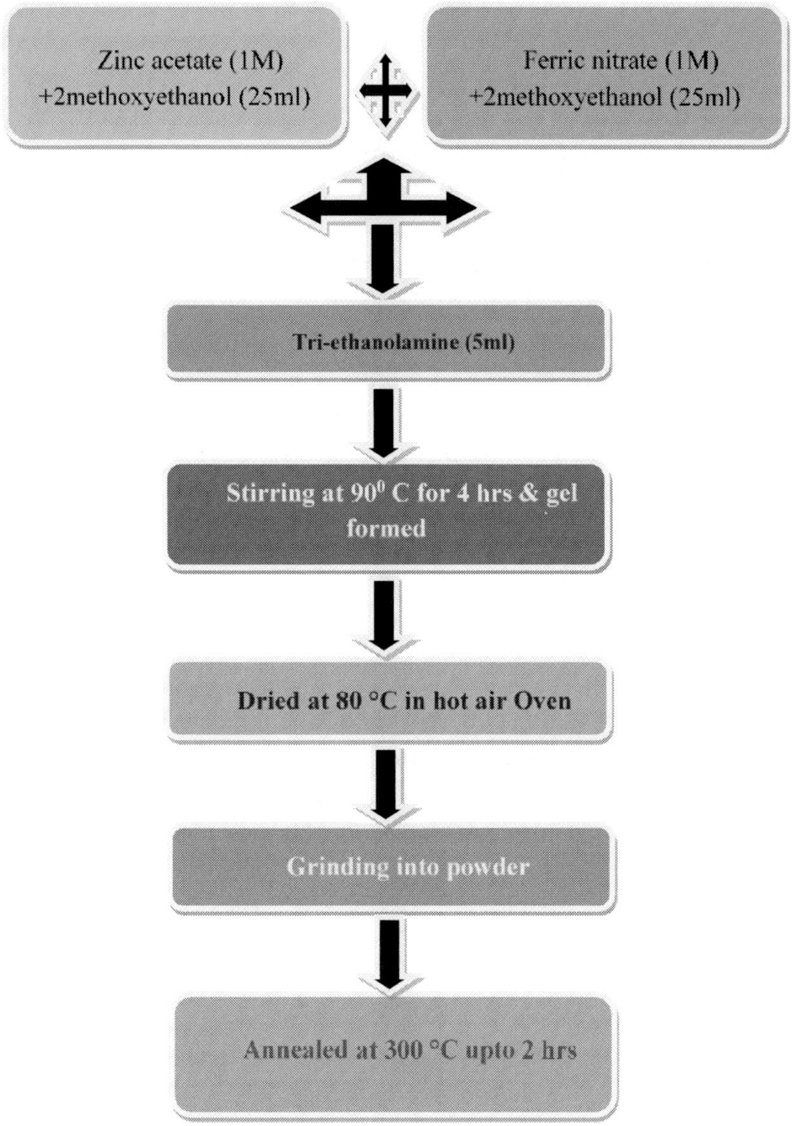

Figure 1. Flow Chart of Fe Doped Zno Nanoparticle.

The Fe doped ZnO nanoparticles have been synthesized via sol-gel method. The analytical grade Zinc Acetate (1M), ferric nitrate (1M), 2Methoxyethanal and tri-ethanolamine (TEA) were used as precursor material. The required amount of zinc Acetate was first dissolved in 25 ml of 2methoxyethanal solution, Further the resultant solution stirring 10 minutes and similarly, the ferric nitrate was prepared in 25 ml of 2methoxyethanal. The two solutions are mixed with constant stirring. The 5 ml of tri-ethanolamine (TEA) mixed with constant stirring solution. The solution was evaporated by constant stirring for 4 hours at 90°C and the gel was formed. Then the gel was dried at 80°C for 2 hrs in an oven. The dried gel was ground to make fine powder. In the same process, the Zinc acetate and Ferric Nitrate with 1% percentage (for the ratios like 99+1%) were prepared. Finally, the products were annealed at 300°C for 2 hrs using muffle furnace. The synthesis samples were analyzed through various characterization techniques XRD, UV-Vis, SEM with EDAX, FTIR, and VSM. The flow chart of Fe doped ZnO is shown in Figure 1.

2.1. Characterizations Techniques

The characterization study of Fe doped ZnO nanoparticles were done to the investigative size, shape and quantity of particles. Number of technique is used for this purpose, including, XRD (X-Ray diffraction), UV-vis (UV-visible), FT-IR (Fourier Transmission Infrared Spectroscopy), Scanning Electron Microscope (SEM), Energy Dispersive X-ray Spectrometry (EDAX) and Vibrating Sample Magnetometer (VSM).

3. Results and Discussion

3.1. Introduction

The synthesized samples were characterized by adopting different physico chemical methods specifically X-ray diffraction (XRD), Fourier transform infrared (FTIR) spectroscopy, vibrating sample magnetometer (VSM). Fe doped ZnO samples were characterized using the powder X-Ray diffraction measurement, which was carried out at room temperature by using EQUINOX 1000: Benchtop X-Ray diffractometer with monochromatized Cukα radiation (λ = 1.54Å). The FTIR spectrum of the as-prepared sample was recorded, with a 'NICOLET ISF'- FTIR Spectrometer using the KBr-pellet technique at a resolution of 4 cm^{-1} over the range 4000-400 cm^{-1}. The magnetic measurements of the synthesized samples were taken by a vibrating sample magnetometer model LAKESHORE 7410 VSM 155 at room temperature.

3.2. Structural properties

3.2.1. X-Ray Diffraction

XRD patterns of synthesized samples for pure and Fe doped ZnO nanoparticle is shown Figure 2 and 3. All the diffraction peaks were indexed to a pure hexagonal wurtzite structure of ZnO nanoparticles. The diffraction patterns are well matched with JCPDS card no (79-2205). The diffraction peaks at 2θ values 32.23°, 34.68°, 36.60°, 47.83°, 56.87°, 63.19°, 68.10°, and 69.21° can be ascribed to the reflections (100), (002), (101), (102), (110), (103), (112) and (201) planes of the ZnO structure, respectively. No extra peak is observed due to Fe of the synthesized samples, which suggests that the doped Fe atom substitute Zn atoms without changing ZnO wurtzite structure.

The simplest and most widely used method for estimating the crystallite size is from the Full Width at Half Maximum (FWHM in radian) of a diffraction peak by Debye-Scherrer's formula;

$$D = 0.9\lambda/\beta\cos\theta.$$

where, D = crystallite size,

λ = wavelength of the X-rays,

β = FWHM (in radians),

θ = angle of diffraction.

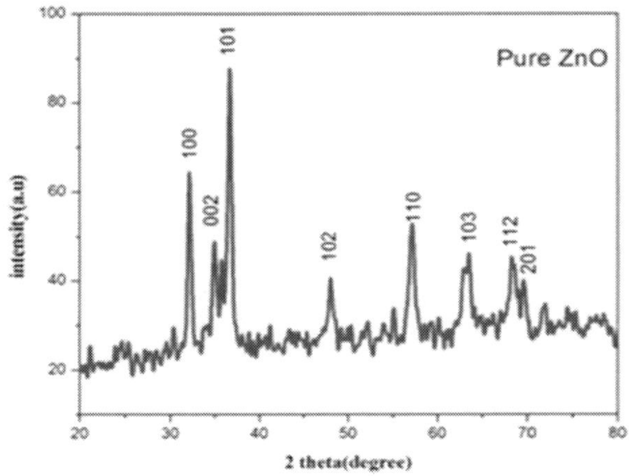

Figure 2. XRD patterns of pure ZnO nanoparticles.

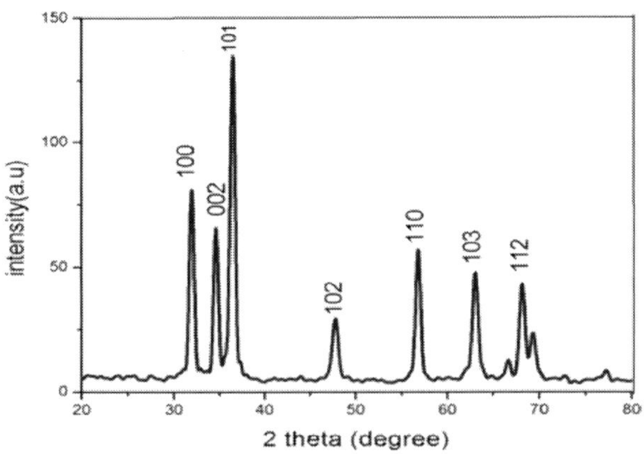

Figure 3. XRD patterns of Fe (1%) doped ZnO nanoparticles.

The size obtained from Scherrer's formula yields the apparent or average particle size for a material. The average crystalline size of the pure ZnO decreased from 30.94 nm to 20.06 nm, when the Fe content doped on the host matrix. The crystallinity of ZnO is restored by the addition of Fe as a dopant which is confirmed by decrease in the intensity of main peak (1 0 1) as the concentration of Fe increased. The average crystallite sizes of the synthesized Fe doped ZnO and Pure ZnO samples are calculated to be 20.06 nm and 30.94 nm, respectively.

3.2.2. Fourier Transform Infra Red Spectroscopy (FTIR)

FTIR is a powerful technique to find out the information about the chemical bonding in a material, identify the elemental constituents of a material sample and supplements the information obtained from XRD and SEM.

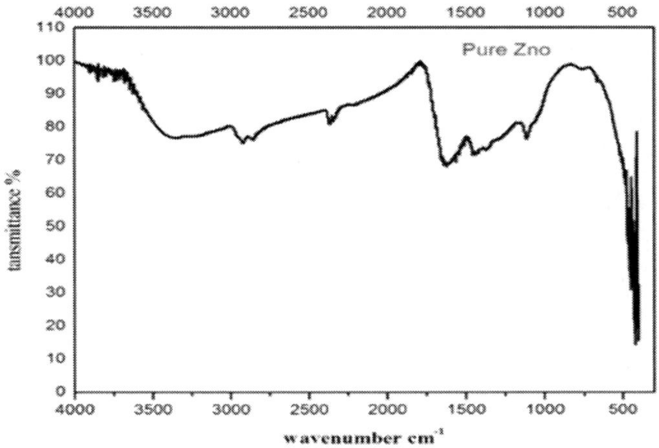

(a)

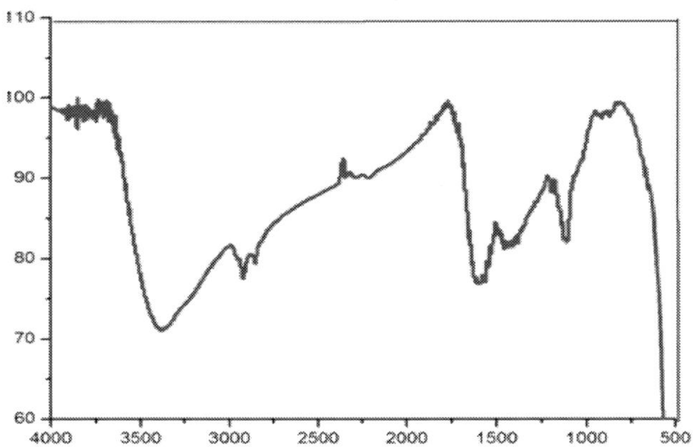

(b)

Figure 4. (a) FTIR spectra of pure ZnO nanoparticles. (b) FTIR spectra of Fe doped ZnO nanoparticles.

FTIR spectra were recorded in solid phase using KBr pellet technique in the wave number region 4000-400 cm^{-1} for Pure ZnO and Fe doped ZnO samples are shown in Figure 4 a and b. This spectral region is very important because of several stretch modes involving hydroxyl bond, carbon to oxygen and metal oxide bonds are obtained clearly in this range. In the FT-IR spectra shown in Figure 4 a and b, the broad absorption band at 3374-3418 cm^{-1} corresponds to the O-H stretching vibration of water present in the sample and the other transmission band at 2916-2932 cm^{-1} is assigned to a residual organic component like C-H streaching mode. The band at 1609-1633 cm^{-1} can be associated with the bending vibration of H_2O molecules. A small peak at 1105-1103 cm^{-1} is corresponding to C-O streaching vibration mode. A strong absorption band informing about the vibrational properties of ZnO is observed for each sample at 497-525 cm^{-1} which is assigned to the Zn-O stretching band which is consistent with that reported elsewhere [6].

3.2.3. Vibrating Sample Magnetometer (VSM)

The magnetic hysteresis loops of the pure and Fe doped ZnO is shown in Figure 5 a and b. Figure 5 (a) illustrates the variation of magnetization (M) as a function of applied field (H) of ZnO nanoparticles. This shows that the pure ZnO nanoparticle is ferromagnetic behavior. The M-H curve of the Fe doped ZnO nanoparticles are shown in Figure 5 (b), It can be seen that the Fe doped sample is in paramagnetic behaviors. However the Fe loops are S-shaped. The saturation magnetization (M_s) of the Fe doped ZnO nanoparticles are shown in table 2 for the samples synthesized with different percentages, respectively, the different magnetization values of the samples may due to their different crystallite size. Therefore, the magnetization has a size-depended behavior for the Fe doped ZnO nanoparticles [7]. As the particle size decreases, the transition takes place from multi-domain to a single domain system. The saturation magnetization decreases with decreasing particle size. On the other hand, coercivity increases because it reaches a maximum value at the critical size for single domain. The prepared Fe doped ZnO should be suitable for spintronic devices.

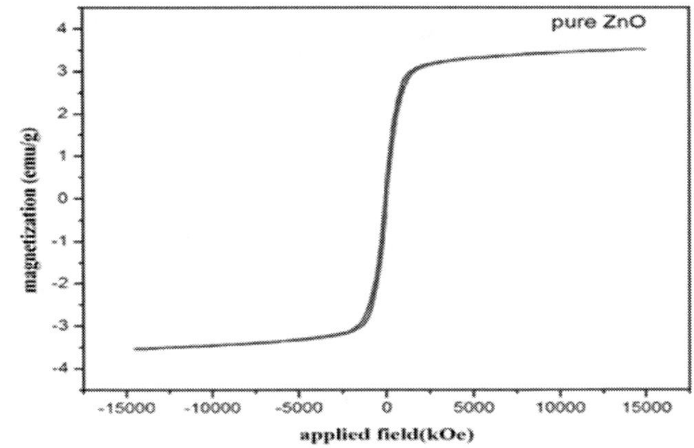

(a)

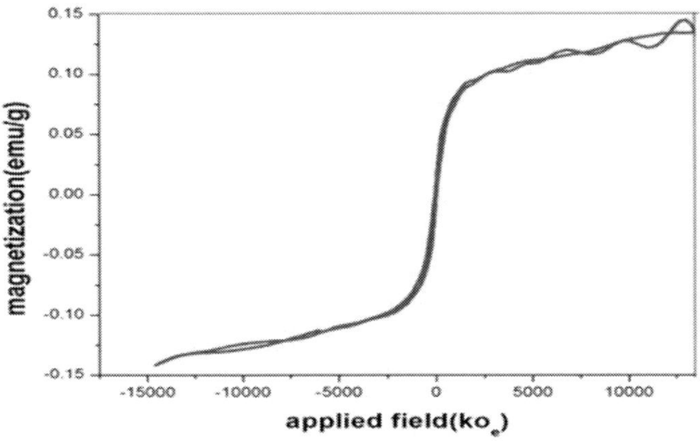

(b)

Figure 5. (a) Room temperature magnetic studies of ZnO nanoparticles. (b) Room temperature VSM analysis of Fe doped ZnO nanoparticles.

Table 2. Magnetic parameters of Pure and Fe doped ZnO nanoparticles

Sample name	Coercivity (Hci) G	Magnetization (Ms) emu/g	Retentivity (Mr) emu/g
ZnO	57.660	3.5342	0.26721
1%	48.519	13.791×10^{-3}	498.96×10^{-6}

Conclusion

- Current work deals with the synthesis, structural, morphology and magnetic properties of pure and Fe doped ZnO nanoparticles by sol-gel method.
- The hexagonal wurtzite structure of the prepared pure and Fe doped ZnO nanoparticles were confirmed by the powder XRD patterns. The average particle sizes of the pure ZnO and Fe doped ZnO nanoparticles are calculated to be 20.06, 24.41, 25.81, 32.29, 22.28, 29.51 nm and 30.94 nm, respectively. It was found that the doping percentages of quantity played important roles in the particle size effect of the Fe doped ZnO nanoparticles.
- The presence of functional groups of synthesized ZnO and Fe doped ZnO nanoparticles were confirmed by the FTIR spectra.
- The magnetic properties of the samples were confirmed by VSM. The results confirm pure ZnO nanoparticle is ferromagnetic behaviour and the M-H curve of the Fe doped ZnO nanoparticles with 1% percentage like are in paramagnetic behavior. The loops are S-shaped, which indicate the ferromagnetic nature of the samples and it is converted into paramagnetic behavior. The different magnetization values of the samples may due to their different particle size. As the particle size decrease, the transition takes place from multi-domain to a single domain system. The saturation magnetization decreases with decreasing particle size. On the other hand, coercivity increases because it reaches a maximum value at the critical size for single domain.
- It has been found that 1% percentage doping has significant effect on the size and magnetic properties of Fe doped ZnO nanoparticles. The prepared Fe doped ZnO nanoparticles show applications in the areas of nano electronics, sensors, batteries and photocatalysis.

REFERENCES

[1] Nair, R., Varghese, S. H., Nair, B. G., Maekawa, T., Yoshida, Y., Kumar, D. S., *Plant Sci.* 179, (2010)154e163.

[2] Osaka, T., Matsunaga, T., Nakanishi, T., Arakaki, A., Niwa, D., Iida, H., *Analytical and Bioanalytical Chemistry* 384: (2006)593-600.

[3] Singh, M., Singh, S., Prasad, S., Gambhir, I. S., *Digest Journal of Nanomaterials and Biostructures* 3: (2008)115-122.

[4] Sinha, S. S., Ieshita Pan, Pompee Chanda and Sukanta K Sen, *Journal of Applied Biosciences* 19: (2009)1113-1130.

[5] Sharma, V. K., Yngard, R. A., Lin, Y., *Adv. Colloid Interface Sci.* 145, (2009)83-96.

[6] Mishra, A. K. D. Das, Investigation on Fe-doped ZnO nanostructures prepared by a chemical route, *Materials Science and Engineering B* 171 (2010) 5-10.

[7] Saha, S., A. K. Bhunia, *Journal of Physical Sciences,* 17, (2013) 191-195.

EDITORS' CONTACT INFORMATION

S. Kannadhasan

Assistant Professor,
Department of Electronics and Communication Engineering,
Cheran College of Engineering,
K. Paramathi, Karur, Tamilnadu, India
E-mail: kannadhasan.ece@gmail.com

R. Nagarajan

Professor,
Department of Electrical and Electronics Engineering,
Gnanamani College of Technology,
A.K Samuthiram, Namakkal, Tamilnadu, India
Email: krrajan71@gmail.com

M. Shanmuganantham

Vice Principal,
Department of Electrical and Electronics Engineering,
Tamilnadu Government Polytechnic College,
Madurai, Tamilnadu, India
Email: shammu1968@gmail.com

INDEX

#

3DCNN, xii, 96

A

active learning, 152, 153, 155
air pollution, vii, xvi, 226, 257, 258, 259, 260, 270, 271, 272
algorithm, xii, xvii, 7, 37, 44, 45, 47, 51, 56, 57, 59, 87, 88, 94, 96, 101, 105, 114, 169, 171, 175, 189, 192, 196, 265, 266, 276, 280, 281, 283, 285, 286
anti-reflection coating, xvii, 298, 314, 320
artificial intelligence, ix, 13, 35, 48
automotive application, 252
automotive applications, 252

B

bacterial infection, xvii, 275, 276
batteries, xiv, 116, 123, 126, 127, 131, 163, 164, 167, 169, 188, 190, 192, 227, 240, 241, 242, 253, 254, 293, 332
bio imaging, 23
biological activities, 67, 83
biological behavior, 66
blended learning, 152, 154, 156, 159, 160, 161, 162
brain, ix, x, 21, 22, 25, 26, 27, 34, 38, 39, 40
brain activity, 40
brain structure, ix, 21
breathing, xvi, 257, 258, 260, 261
bronchospasm, 260, 261

C

cancer, xvii, 25, 27, 28, 29, 31, 197, 275, 276
carbon monoxide, 230, 250, 262, 271, 272
catalyst, 66, 227, 229, 234, 238, 239, 244, 250
chemical characteristics, 188
chemical inertness, 310
chemical properties, 312
chemical reactions, 245
chemical stability, 309, 311

chemotherapeutic agent, 67
chronic obstructive pulmonary disease, 258, 259
classification, xii, xv, 45, 48, 49, 51, 56, 57, 60, 61, 96, 101, 104, 105, 109, 111, 114, 170, 195, 196, 206, 266, 281, 284, 285
classroom, xiv, 152, 154, 155, 159, 160, 161, 162
cognitive neurosciences, ix, 21, 23
commercial, xiii, 76, 115, 126, 131, 187, 251, 256, 300
communication, xii, xiv, 36, 95, 96, 97, 98, 99, 151, 155
communication skills, 155
communication systems, 36
community, ix, 116, 120, 122, 127, 153, 158
community energy storage, vi, ix, xiii, 115, 116, 119, 120, 122, 126, 127, 131, 132
comparative analysis, xi, 42, 207, 209
complexity, 137, 148, 186, 282, 283
compounds, xii, 66, 72, 74, 76, 81, 231, 303
computation, 35, 53, 87, 101, 104, 109
computer, ix, xii, 2, 7, 23, 24, 25, 35, 36, 82, 95, 96, 97, 136, 152, 156, 173, 263
concurrent K-means, xvii, 275, 280
conduction, 218, 237, 299, 302, 303, 307
configuration, 100, 106, 148
converter, vi, xiii, xv, 135, 136, 137, 146, 148, 149, 150, 213, 214, 215, 216, 217, 220, 221, 222
coordination, 97, 303, 307, 309
cost, xii, xiii, xvii, 34, 85, 86, 87, 88, 89, 90, 91, 92, 109, 119, 127, 131, 135, 153, 171, 173, 197, 211, 227, 239, 242, 243, 244, 245, 246, 248, 249, 250, 255, 284, 287, 290, 297, 304, 314
cycles, 11, 47, 119, 127, 173, 174, 179, 184, 186, 309

D

data, iv, v, vi, xi, xiii, xv, xvi, 2, 6, 7, 8, 10, 11, 14, 16, 34, 35, 36, 37, 41, 42, 48, 52, 56, 57, 60, 61, 81, 88, 96, 97, 99, 100, 101, 106, 107, 112, 158, 163, 164, 165, 166, 170, 171, 172, 173, 175, 176, 177, 179, 180, 181, 184, 185, 186, 187, 189, 190, 192, 195, 196, 197, 206, 210, 211, 257, 258, 259, 262, 265, 271, 272, 273, 280, 281, 302, 309, 322
data mining, xv, 195, 196, 210
data processing, 172, 181
data set, 36, 197, 280
data structure, 173
database, 7, 259, 263, 264
Dc/Dc converter, 136
DC-CNN, xii, 96, 108, 111, 114
deep learning, 34, 56, 62, 105, 113, 178, 277
detection, ix, xii, xvii, 58, 60, 95, 96, 100, 101, 166, 275, 276, 278, 281, 285
detection techniques, 166
diabetes, xv, 195, 196, 206, 210, 258
diabetic patients, 210
diabetic risk, vi, ix, xv, 195, 196, 198, 206, 208, 209
diffusion, x, xii, 22, 32, 66, 71, 74, 76, 229, 238, 310, 320
diseases, x, xvii, 22, 23, 32, 196, 260, 275, 277, 280, 282, 283, 285
distribution, 42, 116, 117, 122, 123, 125, 136, 226, 245, 246, 251, 291, 324

E

education, xiv, 152, 155, 156, 158, 159, 160, 161
educational programs, 157
educational settings, 155, 156
E-learning, vi, 151, 152, 159
electric current, 289
electricity, xvi, xvii, 42, 124, 126, 226, 227, 228, 238, 240, 287, 288, 289, 290, 291, 292, 293, 294, 298, 300, 312
electrodeposition, 191
electrodes, 227, 230, 233, 237, 238, 245, 302, 304, 310
electroencephalography, 38
electrolysis, 234, 236, 239, 252, 253, 254
electrolyte, 227, 228, 230, 232, 233, 235, 236, 237, 238, 239, 299, 302, 304, 305, 306, 310, 311, 314
electromagnetic, 30, 32, 314
electromagnetic fields, 32
electron, 69, 227, 238, 299, 300, 303, 305, 309, 312
electronic circuits, 136, 324
electronic systems, 136
electrons, 230, 238, 289, 291, 299, 300, 302, 307, 311
energy density, 119, 164, 240, 241
energy input, 298
energy supply, 117
engineering, ix, 1, 2, 23, 55, 88, 170, 190, 322
environmental change, 293
environmental conditions, 180
environmental issues, 128, 226
environmental sustainability, 310
equipment, 116, 123, 136, 152, 164, 244, 245, 270, 292
ethanol, 77, 78, 79, 80, 81
execution, 12, 48, 51, 52, 54, 55, 108, 111, 259, 263, 283
exposure, 5, 25, 27, 38, 261, 271, 272
extraction, 49, 99, 101, 280, 282, 284
extracts, 101, 276, 280, 284

F

fault detection, 58, 59, 60, 61
flipped learning, 152, 155
force, 48, 49, 54, 97, 292, 303
fuel cell, ix, xiii, xvi, 135, 148, 192, 225, 227, 228, 230, 231, 234, 235, 236, 237, 238, 241, 242, 243, 244, 245, 249, 250, 252, 253, 254, 255, 256
full-bridge rectifier, 136, 137
fuzzy membership, 201, 204, 207
fuzzy set theory, 196, 207, 208, 209, 210

G

gesture detection, xii, 95, 96
grey prediction, xv, 195, 196, 207, 208, 209, 210

H

HCI, xii, 95, 96, 97, 99, 100, 331
health, xvi, 153, 154, 155, 156, 159, 188, 196, 197, 211, 257, 258, 259, 260, 262, 271, 272, 276
heuristics, 2, 8, 9, 10
hibiscus etc., 276
human, ix, x, xii, 1, 2, 4, 5, 12, 14, 17, 21, 22, 35, 38, 39, 40, 83, 95, 96, 97, 99, 110, 197, 261, 262, 266, 271, 276, 285
human brain, x, 21, 38, 39, 40
human condition, 17
hybrid, xi, 42, 62, 82, 125, 148, 158, 165, 186, 187, 192, 236, 252, 254, 256

hydrogen, ix, xvi, 32, 226, 227, 229, 231, 232, 233, 234, 235, 238, 240, 241, 242, 243, 244, 245, 246, 247, 248, 249, 250, 251, 252, 253, 254, 255, 256

I

ideal, 3, 8, 14, 52, 158, 171, 175, 302, 306, 310
illumination, 299, 301, 302, 303
image, ix, x, 21, 22, 23, 24, 25, 27, 29, 30, 33, 34, 35, 36, 37, 38, 53, 100, 101, 104, 254, 278, 279, 280, 281, 286
images, xi, xii, 23, 26, 27, 30, 33, 34, 35, 36, 37, 54, 96, 101, 254, 277, 278, 279, 280, 281, 282, 285
indexer, 258, 265
India, xiii, 1, 41, 85, 93, 112, 115, 116, 126, 127, 128, 129, 130, 131, 132, 151, 153, 156, 163, 195, 196, 213, 225, 257, 258, 275, 276, 287, 288, 290, 293, 297, 321
individuals, 8, 31, 98, 261, 262
inductor, 137, 141, 142, 215, 216, 217
infrastructure, xvi, 226, 243, 245, 251, 255, 256, 259
integration, xiii, 43, 115, 118, 125, 126, 131, 132, 306, 319, 324
intelligence, 2, 4, 5, 11, 34, 192
interface, xii, xiv, 6, 7, 96, 99, 136, 221
international scenario, vi, 115, 116, 119
IR, xii, 65, 66, 67, 76, 100, 259, 299, 303, 304, 308, 326, 330
issues, xi, xvi, 2, 4, 7, 8, 9, 10, 11, 12, 14, 42, 51, 54, 119, 120, 124, 210, 232, 243, 244, 254, 258, 270, 288
iteration, 171, 173, 176, 184, 185

K

KCA, xii, 96
K-curvature, xiii, 96
ketone, v, 65, 66, 67, 68, 69, 70, 72, 74, 77, 78, 79, 80, 81, 83
ketones, xi, 65, 67, 68, 70, 71, 72, 73, 74, 75, 76
keyword search, vii, xvi, 257, 258, 259, 263, 267, 273
Kirby-Bauer disc diffusion method, 66, 76

L

learners, xiv, 151, 152, 153, 154, 155, 156, 281
learning, ix, xiv, 2, 3, 4, 6, 8, 11, 14, 15, 16, 34, 45, 49, 51, 61, 62, 96, 108, 109, 110, 111, 112, 151, 152, 153, 154, 155, 156, 157, 158, 159, 160, 161, 170, 175, 176, 179, 180, 184, 186, 187, 190, 193, 197, 266
light, xiii, xvii, 48, 96, 98, 100, 180, 289, 290, 291, 292, 293, 298, 301, 302, 304, 306, 308, 309, 313, 314, 316, 319
light conditions, xiii, 96
light scattering, 306
light transmittance, 304
lithium, 123, 127, 169, 172, 174, 179, 188, 191, 192, 232, 237, 253, 254

M

machine learning, xii, 56, 61, 62, 95, 165, 186, 190
magnetic properties, 322, 323, 324, 332
magnetic resonance, x, 22, 25, 32, 35, 38, 39, 40
magnetic resonance (MR), x, 22

magnetic resonance imaging, 25, 35, 38, 39, 40
magnetic resonance spectroscopy, 40
mass, 152, 244, 246, 271, 304, 314, 324
matter, iv, x, 22, 27, 37, 261, 271, 291
measurement, 53, 54, 156, 164, 166, 168, 169, 267, 327
medical, ix, x, xvi, 22, 23, 24, 33, 34, 35, 36, 37, 161, 196, 197, 210, 257, 258, 259, 260, 262, 263, 269, 276
medical image processing, v, ix, x, 22, 23
medical keywords, 258
medicine, 23, 24, 32, 33, 196, 210, 211, 285
membranes, xvi, 226, 234, 236, 237, 239, 245, 254
meta-analysis, 154, 155, 156, 158, 159, 160, 272, 284
methodology, xii, 14, 16, 51, 85, 98, 99, 191, 259
mining, xv, 2, 16, 48, 60, 61, 195, 196, 197, 210
M-learning, xiv, 152, 156
models, 10, 11, 15, 34, 45, 49, 52, 53, 56, 88, 108, 166, 179, 184, 187, 193, 196, 207, 208, 209, 210
motivation, 42, 98, 154, 155

N

nanoparticles, xviii, 321, 322, 324, 326, 327, 328, 329, 330, 331, 332
national scenario, 116, 126, 249
neural network, xii, 46, 48, 61, 62, 96, 113, 114, 173, 175, 176, 177, 178, 179, 186, 190, 191, 193, 285
neuroimaging, x, 22, 23, 31, 33, 40
neurons, 3, 12, 14, 27, 56, 105, 110, 172, 173, 175, 178, 186
nitrogen dioxide, 260, 271

O

online learning, xiv, 151, 152, 154
online Learning, 152, 153
ontology, 2, 8
operations, 27, 175, 180, 187
opportunities, x, 21, 126, 157
optimization, xi, 2, 9, 12, 37, 42, 45, 61, 62, 87, 92, 93, 170, 171, 184, 190, 306, 320
oxidation, 229, 231, 232, 233, 234, 305, 322
oxygen, 227, 229, 231, 232, 233, 235, 238, 250, 262, 271, 304, 330
ozone, 258, 260, 261, 271

P

photons, xvii, 289, 298, 299, 315
photosensitizers, 309
plant disease, 276, 285
plants, 86, 120, 123, 126, 281, 289, 290, 292
platinum, 227, 234, 238, 244, 250, 255, 311
policy, xiii, xvii, 116, 128, 131, 184, 201, 287, 288, 292
policy instruments, xvii, 287, 288
polling, 52, 54, 104
pollutants, 188, 298
pollution, xvi, 164, 226, 257, 258, 259, 260, 262, 270, 271, 272, 288, 290
polymer, 172, 233, 235, 237, 238, 305
polymer electrolytes, 237
power factor correction, xiii, xv, 115, 125, 131, 213, 214, 215, 220, 221, 222
power generation, xii, 85, 89, 90, 92, 238, 251, 289, 304
power plants, xi, xiv, 42, 163, 252, 289, 292, 293

preparation, iv, xviii, 15, 47, 49, 52, 55, 98, 108, 112, 305, 306, 322
programming, ix, 1, 2, 7, 15, 52, 198, 266
project, 100, 116, 123, 124, 125, 127, 231, 288
pulsating current cell, 136
pulse discharge, 181
purchasing power, xvii, 288

Q

quantification, xiv, 163, 186
query, 258, 259, 260, 262, 265, 267
query search, 258

R

radiation, 25, 27, 28, 29, 30, 31, 38, 291, 292, 304, 307, 314, 327
real time, xiv, xv, 29, 30, 44, 53, 54, 57, 102, 164, 165, 166, 167, 169, 170, 172, 180, 181, 186, 187, 195, 210
recognition, xii, 11, 12, 24, 47, 53, 61, 96, 100, 101, 102, 103, 105, 106, 107, 110, 112, 113, 114, 155, 277
rectifier, xv, 137, 213, 214, 220, 221
redundancy, 203, 265, 267, 268
rehabilitation, x, xi, 21, 22, 23, 39
reliability, 116, 117, 121, 125, 127, 243, 244, 245, 255
renewable energy, xi, xiii, 42, 58, 115, 118, 125, 126, 132, 135, 136, 148, 226, 238, 251, 252, 289, 290, 291, 294, 312
renewable energy technologies, 290
requirement, 3, 117, 131, 244, 298, 305
review, v, vi, x, 3, 21, 39, 41, 45, 57, 58, 82, 115, 116, 132, 154, 155, 158, 159, 160, 161, 162, 163, 188, 196, 225, 250, 253, 254, 255, 271, 272, 284
reweighted KNN linear, xvii, 276, 281, 284
ripple, xv, 213, 214, 217
risk, ix, xv, 6, 46, 195, 196, 197, 206, 208, 209, 211, 238, 260, 276
rough set theory, xv, 195, 196, 207, 208, 209
rules, 184, 203, 205, 206, 207, 209

S

science, ix, x, 6, 12, 22, 189, 190, 322
segmentation, 24, 35, 36, 101, 276, 277, 280, 285
semiconductor, 299, 303, 305, 312, 324
services, iv, xiii, 115, 116, 117, 127, 131
set theory, xv, 195, 196, 207, 208, 209
SIFT, xiii, 96
society, xvi, 97, 211, 257, 293
sodium, 121, 123, 231, 232
soft set theory, xv, 195, 196, 208
software, ix, 1, 2, 33, 34, 47, 88, 221
solar cells, ix, xvii, 291, 292, 297, 298, 302, 304, 314, 319, 320
solar energy, vii, xvii, 118, 127, 287, 288, 289, 290, 291, 293, 294, 298, 299, 312, 324
solar energy project and solar photovoltaic, 288
solution, xiii, 87, 88, 124, 135, 154, 157, 187, 299, 303, 306, 319, 326
stability, 42, 58, 86, 117, 125, 127, 185, 288, 299, 309
storage, ix, xiii, 33, 37, 115, 116, 118, 120, 121, 122, 123, 124, 125, 126, 127, 131, 132, 136, 164, 191, 242, 243, 249, 252, 253, 254, 289, 292, 322
stress, 97, 125, 191, 270
stretching, 37, 67, 330
stroke, 40, 62, 258, 270
structure, xiii, 7, 14, 15, 47, 57, 82, 101, 105, 108, 110, 135, 148, 169, 180,

229, 236, 304, 315, 323, 324, 327, 332
substrate, xvii, 298, 303, 304, 305, 306, 307, 314, 316, 318
superlative set theory, vi, xv, 195, 196
synthesis, xi, xviii, 65, 66, 82, 83, 157, 158, 159, 316, 322, 324, 326, 332

T

techniques, ix, x, xi, xii, xiv, xvi, xvii, xviii, 11, 12, 21, 22, 23, 32, 36, 37, 39, 42, 51, 57, 61, 62, 86, 95, 100, 102, 106, 164, 165, 166, 169, 170, 171, 177, 178, 181, 186, 187, 196, 197, 214, 221, 246, 251, 260, 263, 275, 305, 322, 326
technological advancement, 244
temperature, xvii, 119, 164, 166, 167, 171, 173, 178, 179, 180, 227, 229, 230, 231, 233, 234, 235, 239, 244, 275, 276, 304, 305, 327, 331
titania, 300, 302, 303, 304, 305, 309
total harmonic distortion, 214, 220
training, 45, 48, 57, 154, 155, 157, 173, 177, 179, 180, 184, 186, 281
transmission, xi, xii, xvii, 37, 42, 43, 47, 52, 61, 62, 85, 86, 87, 88, 89, 90, 91, 92, 117, 123, 126, 136, 291, 298, 315, 330
transport, 50, 237, 246, 249, 302, 305, 306, 310, 314, 322
transportation, 116, 131, 227, 236, 237, 239, 243, 246, 250, 252
treatment, x, xi, xvii, 21, 22, 23, 26, 30, 35, 197, 258, 262, 275, 276, 284, 306
trial, 154, 161, 205, 284

V

variables, 98, 170, 179, 200, 201
vector, 12, 47, 49, 59, 60, 61, 62, 105, 173, 190, 277, 285
vehicles, xiii, xiv, xvi, 116, 125, 135, 148, 163, 225, 227, 240, 241, 242, 243, 244, 249, 250, 252, 254, 255, 256, 293

W

water, 38, 39, 116, 131, 227, 229, 230, 234, 237, 238, 244, 245, 250, 251, 254, 261, 270, 289, 290, 294, 304, 309, 330
web, xvi, 153, 258, 262, 272
weight loss, xvii, 275, 276
worldwide, 120, 123, 152, 156

Z

zinc, 323, 326
ZnO, vii, xviii, 297, 299, 305, 312, 313, 314, 315, 316, 317, 318, 319, 320, 321, 324, 326, 327, 328, 329, 330, 331, 332, 333
ZnO nanostructures, 324, 333